# *Cases in Technical Communication*

# *Cases in Technical Communication*

**Andrea Breemer Frantz**
*Buena Vista University*
*Storm Lake, Iowa*

**Wadsworth Publishing Company**
I(T)P® An International Thomson Publishing Company

*Belmont, CA • Albany, NY • Boston • Cincinnati • Johannesburg • London • Madrid • Melbourne*
*Mexico City • New York • Pacific Grove, CA • Scottsdale, AZ • Singapore • Tokyo • Toronto*

Editor: Karen Allanson
Editorial Assistant: Godwin Chu
Print Buyer: Barbara Britton
Production: Vicki Moran
Copy Editor: Victoria Nelson
Illustrator: Carl Yoshihara
Cover Design: Cuttriss & Hambleton
Cover Image: Leonardo da Vinci, study of a flying machine, *Codex* B, vol. 80r. Paris, Bibliotheque de l'Institut de France. Photo: Scala/Art Resource, NY.
Compositor: Publishing Support Services
Printer: Webcom Limited



Printed in Canada
1 2 3 4 5 6 7 8 9 10

For more information, contact Wadsworth Publishing Company, 10 Davis Drive, Belmont, CA 94002, or electronically at http://www.wadsworth.com

International Thomson Publishing Europe
Berkshire House
168-173 High Holborn
London, WC1V 7AA, United Kingdom

Nelson ITP, Australia
102 Dodds Street
South Melbourne
Victoria 3205 Australia

Nelson Canada
1120 Birchmount Road
Scarborough, Ontario
Canada M1K 5G4

International Thomson Publishing Southern Africa
Building 18, Constantia Square
138 Sixteenth Road, P.O. Box 2459
Halfway House, 1685 South Africa

International Thomson Editores
Seneca, 53
Colonia Polanco
11560 México D.F. México

International Thomson Publishing
Asia
60 Albert Street
#15-01 Albert Complex
Singapore 189969

International Thomson Publishing
Japan
Hirakawa-cho Kyowa Building, 3F
2-2-1 Hirakawa-cho, Chiyoda-ku
Tokyo 102 Japan



ISBN 0-534-51507-X

*This book is printed on acid-free recycled paper.*

*In Loving Dedication to Michael and Hannah—*
*You seem to know when and when not to ask how work is coming.*
*Thanks for your intuitive love and support.*
*They mean more than I can ever put into words.*

# *Contents in Brief*

# Contents in Detail

## PART II Developing the Communicator's Tools 47

# *Preface*

## To the Student

Welcome to the world of technical writing...well, sort of. This book is designed to introduce you to a wide array of workplace writing scenarios that require you to create technical documents of all kinds. While the cases offered here are based on real-world communication issues in actual organizations, you and I both know that you're not actually going to produce material from these cases that will really be used in the organizations themselves. Or will it?

Case-based writing has long been valued as one of the most realistic means of encouraging students to experiment with communication strategies that they will ultimately use in eventual workplace settings. Aside from actually working with a client and producing public documents, responding to a case is the most comprehensive method you may have of dissecting contextual information, analyzing the complexities of a multilayered audience, and organizing and writing about complex technical information. So, while the memos, reports, and designs you create in response to these cases may not be returned to the architectural firms, manufacturers, environmental corporations, or hospitals you produce them for, the *skills* you gain will indeed follow you into the professional positions you take once you leave school.

But let's back up for a second. For some of you, these will be the first cases you've ever seen. What are cases, exactly? In a nutshell, they are like interactive stories that ask you to produce something in order to move to the next step—you may certainly think of them as you might a role-playing game. You are a character within the narrative faced with a specific communication challenge. In some of these cases, those challenges may be relatively uncomplicated. In others, you may be faced with several barriers blocking your way to a simple response. You are therefore asked to draw on all of your communication skills—sometimes alone, sometimes with collaborative partners from your class—and respond in the most logical, professional way you can. Your response, however, as in real life, will only add to the many layers of organizational communication. Therefore, it is important that you see your response—be it a memo, a report, or a set of instructions—as *one step*, not necessarily the *final* step in the flow of communication within the organization. Whatever you produce will undoubtedly produce a response. Communication—oral, written, visual—responds to and creates a ripple effect. That's why cases are actually so similar to role-playing games. Technically they can go on for as long as you'd like them to.

Because little in life is cut and dried, you will find that your classmates will more often than not respond to the situation differently than you do. Let's face it: you're an individual with unique experiences and tastes; therefore, the details in the case that are most attractive or puzzling to you may be details others ignore or respond to in ways that surprise you. In addition, the situations offered in the cases, though they may have happened in specific organizations, will probably not occur again in the same way in another organization. The particulars will always change, and that's what makes writing such a constant challenge—and interesting. What these cases *will* guarantee, however, is that the practice they offer you in analysis, research, and organizational strategies will ultimately be an invaluable warm-up for the real thing.

You should have the opportunity in class to discuss the details of the case and perhaps even to brainstorm for possible solutions. While none of these cases expects you to be a professional engineer, marketing specialist, or expert in chemical compounds, they may introduce you to terminology with which you are unfamiliar or sets of data you have never seen before. Don't worry about your lack of expertise there. Your job is to be able to decipher and communicate about the details. However, a successful response to a case often is dependent not only on how skillfully a writer articulates a message, but also on how much initiative he or she has taken to get to that point. For example, if you are faced with a word you don't know, don't hesitate to look it up. If you are unclear about a technical process you are trying to describe, test it out or ask someone who might know. Good writers use as many resources as they can to get their message across most effectively. Ultimately, your responses to the majority of these cases can only be enhanced if you take the initiative to go beyond the background information offered here. Use your library. Use the experts in town. Use the 800 numbers on the backs of products. Don't be afraid to push the boundaries of what is offered here to learn more about your subjects.

The best advice I can give you about how to respond to the cases in this book, however, is to relax and have fun with your projects. These cases will allow you to stretch your creative muscles and experiment with ideas you've likely never considered before. Take risks. Develop details. Enjoy the freedom you have here to try things without the risk of professional failure. Then take what you've learned into the work force and engage in careful, conscientious, and ethical communication practices.

## To the Technical Communication Instructor

How many of us have faced a room full of frustrated students who only want a straightforward and honest answer to the question, "How is this assignment going to help me once I leave here?" It's a tough question (and the students know it) because it asks teachers to be fortunetellers, to some degree. How can I be sure, for example, that Hannah, a community and regional planning student, will have any use for a rhetorical analysis of a set of technical instructions? What guarantees that Chris, a chemical engineering student, will ever use the skills she gains in analyzing a request for proposals?

As with most elements of teaching, there are no guarantees. No one assignment—in *any* class—can meet the needs and expectations of all students. However, this doesn't negate the importance of the students' question; nor does it mean that we should patronize them with answers like, "You will all be required to write a variety of things when you enter the work force. The more practice you get here, the better off you'll be." Such an answer rings of "because it's good for you," and although it is not a dishonest answer per se, it feels somehow dissatisfying because it doesn't get at the heart of the matter. Students want to understand why *this* writing task? Why *this* way? The question begs a more particularized response, and that is what this books attempts to do.

### Why a Casebook?

One of the ways I have attempted to answer students' needs for concrete purpose in their writing tasks is to offer them detailed cases that ask them to directly address a wide variety of issues and invoke an array of skills. Most students are not employed full time in professional positions as they attempt to complete their undergraduate degrees. As a result, students have little access to firsthand workplace scenarios that come from personal experience. I have found that cases based on actual workplace scenarios address students' desire for concrete purpose in their writing tasks while also offering them invaluable insight into workplace writing contexts.

While upper-level students develop technical skills of presentation relatively quickly (e.g., mechanical correctness or basic organizational development), critical thinking skills for analyzing and responding to specific contextual ambiguities (e.g., discord and struggle between divisions of a company over a communication process) are like muscles; they develop more slowly, and only after much exercise. Detailed cases offer insight into all elements of workplace writing, including less tangible, contextual issues like power relations, time constraints, risk taking, and conflict, to name a few. Therefore, one of the key purposes for employing communication cases is to offer students insight into the real contextual elements that make up the sorts of writing scenarios they are likely to face. While they may not face cases exactly like the ones offered in this book, these scenarios offer access to realistic issues students will learn from by applying critical thinking skills.

Cases also offer students the advantage of learning and experimenting with professional communication skills in the relatively *safe* environment of the classroom. For example, if an entry-level professional chooses to ignore a supervisor's recommendation to take out a crucial section in an annual report in order to save space, the writer's choice could cost the company money, time, and potentially, in a worst-case scenario, her job. While the writer's choice may indeed be the right one for the sake of ethics, design, or the integrity of the organization, to contradict a superior rarely goes without consequences. Communication cases like those offered in this book offer students the opportunity to make such crucial choices with less professional risk. While students still perceive grades and reputation among their peers as "risks," most will also agree that they prefer to experiment with those risks prior to their entry into the workplace. Cases allow students to test a variety of responses as well as to see how others have answered the same situation. The latter is important because it underscores the complexities of organizational contexts and the fact that a variety of approaches may, in fact, be useful and appropriate.

The cases in this book offer insight into a wide array of workplace situations from various professional vantage points. Each scenario asks students to carefully examine contextual information as they solve communication problems. While the information offered in the cases may stand alone as the sole basis for a student's written response, this does not preclude students supplementing the response with research of their own initiative. One of the philosophies grounding this text is that successful professional communication is highly dependent upon individual initiative, which may in some cases mean moving beyond what is provided.

## How Do I Use This Casebook?

Because each teacher and class has different text needs, this casebook is specifically structured and organized to encourage a wide variety of uses. Thus, it's useful here to examine the organization of the book, ways of reading and using the cases, and possible evaluation strategies.

**Organization** Although this casebook may certainly be used alone for more advanced students, its structure—based in part on Rebecca Burnett's best-selling *Technical Communication* text—depends on detailed background development with the help of a more comprehensive textbook. It is important, for example, that before students are asked to chart statistical information in Case 9 or develop a detailed oral report in Case 4, they have the insight a technical communication text will provide. Though I have used Burnett's *Technical Communication* as my own guide to structuring and developing these cases, I believe the cases can stand on their own or work alongside most other technical communication texts on the market.

Having taught a variety of professional communication courses, I am the first to admit that the character of a given section can easily determine how and if I cover specific elements of the general course. A group made up largely of engineering students, for example, may require more emphasis on report and proposal writing; a class that boasts a wider cross-section of majors may require more intensive examination of contextual constraints and workplace cultures. Consequently, though the order of cases presented here follows what I think is logical development—from understanding contexts and analyzing conflict to completing complex documents and maintaining a comprehensive view of the importance of professional communication practices in the workplace—you need not feel bound to follow the organization as I have patterned it here.

The pattern I offer should instead be understood simply as *one* of many possible patterns that depend upon the nature of the course you offer and the philosophical approach you take in it. One of the most important characteristics of any text is flexibility—understanding that each teacher will reconstruct the order and importance of the cases to fit the particular needs and emphases of the course he or she is teaching. I recognize that there may be some cases that fit the needs and the makeup of your class more than others and that as a result you may feel compelled to either completely ignore some of these cases or add significantly to one or two. The emphasis you place on information and the order in which you offer these cases can— indeed, *should*—be tailored to meet the goals you have set forth for your technical communication students.

**Reading the Cases and Tasks** Most of the cases in this book are multiple-task assignments that ask students to produce letters, memos, reports, feasibility reports, visual designs, and oral presentations, all of which are widely regarded by professionals as standard genres entry-level professionals are expected to be able to produce. While the documents students produce in these cases reflect the genres that technical and professional texts illustrate, the situations out of which the documents are produced are particularized to reflect an individual organizational climate and culture. In addition, you will find some of the tasks ask students to work in teams. Many professionals have acknowledged the importance and widespread trend toward the use of collaboration at all stages of document creation, including data gathering, drafting, assessment, testing, and revision. Accordingly, I have incorporated group work throughout the cases at all stages to reflect workplace climates and the trend toward recognizing and valuing collaborative work.

The cases offered here are based on information provided by numerous professionals from actual organizations and from research; however, the details, characters, and issues highlighted, while based on fact, are fictitious. Because of proprietary issues for some organizations, I have changed some data and details to maintain anonymity.

You'll note that the tasks at the end of each case involve everything from background development and gathering further information to responding to "major" tasks that often involve developing more formal documents or presentations. Although the tasks are designed to build on one another, it is usually not necessary to assign them chronologically (as they appear) or even to assign all of them to any given class. Depending on the needs of the class you are teaching, your students may benefit most from an in-depth discussion of the complexities of a given case as opposed to following through with a written response. The cases are flexible and should be used as they accord with the course and the level of experience students have. Therefore, while the cases offer background information gathering tasks, major tasks, follow-up tasks, and some correspondence or other tasks, you should feel free either to rearrange the order in which they are presented and completed or to choose only the one or two steps pertinent to your class needs. The following list offers a brief summary of the purpose behind each section of cases.

- **Part I: Understanding the Communicator's Work** introduces students to issues of context, audience analysis and collaboration. In the cases offered here, students will experiment with responding to texts and contexts and practice writing for a variety of readers both within and outside the organization in which writers find themselves working.
- **Part II: Developing the Communicator's Tools** introduces students to issues of research and documentation, visual design, and testing and revising texts. The cases in this part ask students to probe complex contexts, research strategies, and the importance of integrated visual and verbal design issues in formal documents.
- **Part III: Applying the Communicator's Techniques** invokes issues surrounding the design of technical descriptions, processes, and abstracts. The cases illustrate technical, legal, and social issues surrounding the design of some of the commonplace forms of communication in the United States.
- **Part IV: Completing Documents** asks students to create sections of longer organizational texts such as reports and proposals. Students will work individually and in teams, employing all of their communication skills as they develop rhetorically rich responses to internal and external demands for organizational accountability and feasibility.
- **Part V: Maintaining Professional Communication** encourages student writers to develop and maintain interactive communication skills in correspondence and oral presentations. Students will work individually and collectively to develop formal and informal presentations and respond to a series of written correspondence.

**Evaluation** One of the toughest tasks a teacher faces is evaluating student work. This task becomes especially difficult as students respond to cases, in part, because the case information is without a doubt less complete than what students would actually have were they producing technical writing or designs that emerged from workplace scenarios in which they really worked. Though evaluation is difficult, we are challenged to respond and to help students make their work better; thus, we are obligated to offer an evaluation system that is both comprehensive and understandable.

I have found that successful responses to student casework often draw together issues of clarity, creativity, attention to detail, interpersonal communication insights, and fine-tuned presentational skills. I offer here a summary of criteria you may use to help guide your responses to the major tasks (and many of the other tasks offered at the end of each case). This summary is designed to serve mainly as a guide for issues and skills you may seek to emphasize as students complete the tasks or discuss in large or small groups. While the summary offers touchstones for evaluation, it purposefully does not place weight or emphasis on its specific components. Clearly, the emphasis placed on specific skills will depend largely on the nature of the case and what it is designed to teach students.

## Summary of Evaluative Criteria for Technical Cases

Although evaluation is guided a great deal by individual experiences and preferences, communication teachers often struggle with students' frustration over the apparently subjective assessments of their work. I have designed an evaluative chart located at the end of each case that may help to structure responses to student work and guide students to a better understanding of the criteria by which they will be judged. The following is a summary of criteria I have found valuable in my own evaluations.

### Purpose/Key Points

- identifies and defines purpose (presented explicitly or implicitly)
- articulates key points
- makes purpose and key points visually identifiable and accessible

### Context

- identifies and responds appropriately to contextual details offered in case background
- demonstrates an understanding of technical information
- articulates technical information accurately and appropriately given the contextual constraints of the case
- can identify technical information that contains inaccuracies or inconsistencies
- demonstrates an awareness of the context of a document, visual, or presentation in an organization
- uses an appropriate design and development process

### Audience

- identifies and defines audience by demographic characteristics, organizational role, receptivity to information, writer's relationship to readers, and other factors
- makes verbal choices appropriate to audience (e.g., organizational hierarchies, ambiguous or diverse audiences, cultural/gender/age issues)

- makes visual choices appropriate to audience (familiar design conventions, etc.)
- presents information that is accessible/comprehensible to the intended audience

**Organization/Development/Support**

- develops a clear line of argument or establishes clear purpose
- collects and uses appropriate support for arguments
- uses an identifiable, appropriate pattern of organization
- demonstrates global and local coherence

**Design**

- demonstrates an awareness of information design
- uses design conventions in ways appropriate to the purpose, audience, and type of document (e.g., external memo, report, or proposal)
- integrates visual and verbal elements in ways appropriate to the purpose, audience, and genre

For those teachers interested in more comprehensive guides to evaluation, I have offered individualized matrixes that break down details at the end of each specific case. Each case matrix highlights issues important to the student's development as a technical communicator. Again, I have avoided placing concrete point value on each issue, since you will likely choose to emphasize a variety of points according to your own expertise and standards. I encourage you to evaluate what you want your students to gain from the case, and place your emphasis in evaluation accordingly.

## Acknowledgments

I don't believe that any book is ever written by a single author. In fact, so many different kinds of contributions have shaped this book that there is no way to completely address everyone who has played a part in its development. I would be remiss, however, if I did not acknowledge some key people without whose insights and expertise the book would never have taken shape at all.

Rebecca Burnett has served not only as the primary reader and person who initiated the idea for this text, but as an invaluable coach and friend throughout the process. Her expertise, insight, and incomparable teaching skills have given life and focus to each of these cases. I also benefited enormously from the anonymous reviewers who looked at early drafts of some of these cases. Their responses were thoughtful, comprehensive, and ultimately helped change the shape of the entire book. Thanks, also, to the Wadsworth editorial and production staff, and in particular to Ryan Veseley for his World Wide Web expertise.

A number of people and institutions provided me with time, technical insights, and access to information that helped the credibility and specificity of these cases. That list includes (but is not limited to) Chris Frantz, Melissa Frantz, Greg Howe, Mary Willig, Ken Hach, Dale Carver, Rod Muilenberg, Darrell Chiavetta, Tim Seydel, the Zond Corporation, and UCAR-OFPS as well as many who have asked to remain anonymous. I greatly appreciate the insights these people and organizations offered and believe they have given the cases authenticity and richness.

Finally, though this is by no means a complete list, I need to thank those who supported and encouraged me throughout the development of this text—my family, my friends Lee-Ann Kastman and Kirstin Cronn-Mills, and most especially my husband Michael, who seemed to always know when *not* to ask about how it was going. Michael's unconditional support and respect for my work is what ultimately helped me finish the project and it is to him, and also my daughter Hannah, that I dedicate this book.


The cases offered here are appropriate for experienced and inexperienced technical communication instructors alike. I hope that they encourage serious discussion and diverse responses among your students. The responses you receive will undoubtedly reflect the uniqueness and intrigue of workplace communication. Good luck!


Andrea Breemer Frantz
Buena Vista University
Storm Lake, Iowa

ITP HIGHER EDUCATION
DISTRIBUTION CENTER
7625 EMPIRE DRIVE
FLORENCE, KY 41042

The enclosed materials are sent to you for your review by
MARY MAJEWSKI

## SALES SUPPORT

| Date | Account | Contact |
|---|---|---|
| 08/28/98 | 676053 | 51 |

SHIP TO: Karla M Kitalong
Michigan Technological Univ
Humanities Department
1400 Townsend Drive
Houghton MI 499310000

## WAREHOUSE INSTRUCTIONS

SLA: 7 BOX: Staple

| LOCATION | QTY | ISBN | AUTHOR/TITLE |
|---|---|---|---|
| K-ASY-086-01 | 1 | 0-534-51507-X | FRANTZ<br>CASES IN TECH WRITING |

INV# 133266473650
PO#
DATE: / /
CARTON: 1 of 1
ID# 8323683

ASSEMBLY
-SLSB

VIA: UP

PAGE 1 OF 1

BATCH: 0637718
030/070

# *Cases in Technical Communication*

# PART I

# *Understanding the Communicator's Work*

# CASE 1

# *Consulting Cap-Hays: Negotiating Technical Texts and Contexts*

This case outlines a scenario in a midsized manufacturing plant that asks you to carefully analyze a section of a technical manual and briefly assess its tone and attention to audience. You have little knowledge of the context for the technical document and must also struggle with a variety of interpersonal issues and pressures connected with the task.

## Background

A turkey manufacturing plant like Cap-Hays has several components to it. The "manufacturing" component is the genetic testing, raising, and breeding of the fowls. Cap-Hays also processes, packages, and delivers turkeys to respective businesses and purchasers.

Cap-Hays is a turkey manufacturing and processing plant that employs just over 400 people in the small town of Ginger, Alabama. In addition to a processing and packaging plant in town, Cap-Hays also manages four turkey farms for breeding and testing. Ginger is a small community of just under 9,000, so Cap-Hays is a major employer, rivaled only by a muffler plant that employs nearly 500 on the edge of town.

For 36 years, Cap-Hays has enjoyed a solid reputation within the community as a good employer with a safe work environment that produces a high-quality product. Cap-Hays employees tend to socialize with one another and think of themselves as part of a large family. At local charity events, Cap-Hays employees almost always offer teams of enthusiastic participants—most of the employees at Cap-Hays are Ginger residents and committed to their community.

Why is it helpful to understand the character of Joseph Capitani here?

Cap-Hays is owned and managed by its cofounder, Joseph Capitani. Capitani's partner, Mack Hays, died in 1990, and the Capitani family bought out the rest of the partnership. Joseph Capitani's children, Angela and Anthony, are also employed by Cap-Hays as business manager and human resources director, respectively. At 65, Joseph Capitani is in no way ready for retirement. An extremely active man, who participates at many levels of his community and his company, he has served on Ginger's city council for the past decade, was the chair for last year's record-breaking United Way fund drive, and is a member of several service organizations and boards. Although Capitani devotes much time to his community, his strongest dedication is to Cap-Hays, which he and Mack Hays built from their combined turkey farms. Capitani has always maintained a strong hand in daily operations at the plant, often filling in for line supervisors when they were sick or on vacation, and consistently refusing to be "stuck behind the desk all of the time." Capitani's official title at Cap-Hays is plant supervisor, and although he is the plant's founder and owner, he is quick to correct people who make the mistake of referring to him as a CEO. "I ain't no corporate type," he reminds people. "I'm just a turkey farmer." Most of his employees call him "Cappy" or "Captain."

Clearly, Cap-Hays has undergone enormous growth and change since the early 1960s. In 1989, for example, Cap-Hays remodeled all four of the largest turkey pens on the north farm and put in new ventilation systems as a result of damage caused by a heat wave of 100-degree temperatures and high humidity. The heat, combined with outdated ventilation systems in the pens, caused Cap-Hays to lose several thousand birds to heat exhaustion, despite attempts to cool the birds with water. Two years later, Capitani ordered the remodeling of the remaining systems on the south farm.

Consider the effects of implementing a labor management system in a business like Cap-Hays. Such a system aims to simplify daily record keeping as well maintain accurate records.

Most recently, Capitani has been considering the addition of new computers and a networked labor management system called Speed Check. Speed Check is designed to replace time clocks and increase middle-management efficiency with attendance/labor data and payroll. Before Capitani agrees to buy and implement the system, however, he has asked Angela and Anthony to look through the marketing materials for Speed Check as well as the training manual and report their impressions about whether it will work for Cap-Hays.

A conflict between Angela and Anthony is described here. What are your impressions of Angela's hesitation to endorse the Speed Check system?

After some time with the materials, the two report back with mixed reactions. While Anthony, the human resources director, believes Speed Check will increase middle-management efficiency with some of the more mundane aspects of their jobs, Angela, the business manager, is not as eager to adopt the new networked labor management system. She cannot fully articulate what her problems with the system are, and she agrees that it holds promise for increasing efficiency; however, there is something about the way the training manual is written that turns her off, and, she adds, it has the potential to turn off other users. While Angela's response is not an all-out condemnation, it gives Capitani some pause before ordering the system.

## The Situation and Your Role

Your role is defined here. What do you anticipate doing in this role?

You are a new communication specialist for Universal Management Systems (UMS). Your job mostly entails overseeing the production of training manuals and making presentations to client groups about a variety of management principles and strategies. Although you are qualified for the position, your job is considered probationary until you have been with the company for six months.

Your direct supervisor, Assistant Director Jody Asher, comes to your office Friday afternoon just as you are sitting down to review a revised draft of a manual for a client. Jody is holding a three-ring binder and says immediately, "I've got a sort of personal favor to ask of you."

Although you have only been on the job three weeks, you like Jody, and, because she is your direct supervisor, you know it is in your best interests to help her out.

"An old family friend just called in a favor and asked me to look at a training manual for a data management system he's considering purchasing. Normally, we'd treat a situation like this as we would any client, but Joseph Capitani is, as I said, an old family friend, and I'm sort of doing this on my own time. I wonder if you'd be willing to use a little bit of your weekend and give me some feedback on something."

"Of course," you reply. "What is it?"

Your purpose is defined here. What is the question Capitani wants answered?

Jody then describes Joseph Capitani's business and the situation as she understands it. "So," she concludes, "Cappy wanted to send the manual to someone he knew who was completely unassociated with Cap-Hays or Ginger to get an honest opinion about this manual. Basically, I think he just really wants to double-check Angela's concerns, even though she can't really put her finger on what it is that bothers her. Now, I really don't know much more than what I've already told you about Cap-Hays, and I don't think he's asking us to tell him whether we think he ought to adopt Speed Check. I think he is fishing more for how this training manual speaks to potential users."

"So, what would you like me to do? Do you want me to analyze the manual?"

There are benefits and drawbacks to examining only a part of the training manual. Benefits include being able to concentrate exclusively on how information in the manual is communicated. Drawbacks include not having all the tools necessary to make a proper assessment of the manual's effectiveness. Discuss other possible drawbacks and advantages.

She shakes her head. "Not the whole thing. Actually, I've been through it already, and I already have some opinions about it. What I really want is to see how another reader, a professional one, responds to how the information is presented here. I'm not looking for your opinions necessarily to *match* mine, of course. I just want to get some gut-level response for your feelings about it. Basically, I just want you to look at a few pages."

"A few pages? Isn't that sort of taking what I read out of context?" you ask, though you're secretly relieved you don't have to read an extra hundred pages of work this weekend.

"Yes, it is taking material out of context, but we're not really concerned about how effective Speed Check is or isn't for Cap-Hays system. Cappy will ultimately make that call. I think what he wants is a professional opinion about how well this technical information is presented in the training manual. Angela's primary concern is that if it doesn't communicate effectively, it will turn off employees who have to use the system before they're even trained. I think Cappy just wants a sense for whether Angela's on the mark or not. In the end, I suppose he's pretty likely to adopt it one way or another, but if the training manual is ineffective, it's best to compensate for that ahead of time."

You nod. It makes sense, but you are still a little uncomfortable with only seeing a few pages of the manual. After all, even though this is a "favor" to Jody, done on your own time, her opinion of your work and your abilities is important to you at UMS. What if your response is completely different from Jody's? Will she then feel inclined to scrutinize your other work more carefully? "I really would like to have the whole manual, though. Just to get a sense for the document."

She shakes her head firmly. "No, I need the whole thing this weekend to make some notes. I promised Cappy I'd get back to him Monday afternoon on

Your task faces some time constraints as well. How might these constraints affect what you can feasibly accomplish?

this. I just figured we could meet Monday morning, you could tell me what you thought about what you read, and I'd use whatever notes you made alongside my own to brief Cappy later. If this is a problem, though, I can ask—"

You cut her off. "No. No. It's no problem. I just thought I could do a more thorough job on it if I had the whole thing. I'll take what you give me and we can go from there."

"Great," Jody says and smiles. "I really appreciate your help on this. Cappy's a good person, and I like helping him out." She pulls out the introduction—as promised, only a few pages—and adds, "Any notes you want to make on this would be helpful, but please don't mark on the actual manual pages. I told Cappy I'd get this back to him intact." You agree to meet at 9 a.m. Monday, and Jody leaves you with your weekend work.

## Background Development

Consider the following questions as a way to fine-tune your understanding of the case and its details. Answer the questions alone or work in small groups of two or three to discuss the answers. Feel free to draw on your responses to inform any of the tasks that follow.

- Jody has simply asked you to provide feedback on what you read. She wants a "gut-level response" to what you see in the manual. At the same time, Cappy has made it clear that he wants a professional eye to look at the manual and respond to it. Is there a conflict in these two directives? If so, can it be resolved? If not, how are the directives complementary?
- Implicit in Jody's request is a sense that you *evaluate* what you read in the manual. You may assume that this means to evaluate the appropriateness of the tone, its attention to audience informational needs, and the effectiveness of design choices. The bottom line is that you are looking for how effectively this portion of the document communicates. Using ideas from chapters 1 and 2 in Rebecca Burnett's *Technical Communication*, 4th edition, create a working list (this is a list that you may add to later) of features or characteristics you believe this manual needs to offer in light of the context you've been introduced to here.
- Using your working list of features or characteristics, choose three or four features you think are *most* important for this technical document. Then discuss with a small group of your peers how you believe you can evaluate those characteristics. For example, if you believe that this manual needs to employ a tone that is both professional and encouraging (because workers using the system will be doing so for the first time and may be intimidated by the computers), how do you judge the tone that you read here? Are the words appropriate to new users who are also professionals who may have worked with Cap-Hays for many years?
- You have only a portion of the manual to examine. What are the possible dangers in examining only part of the text? Can you think of any advantages?

## Getting Started

Examine the pages of the manual's introduction, found in the case appendix, and read them carefully. Briefly analyze what you see the manual's purpose and audience to be in some notes to yourself. In your examination, consider whether the tone established by the manual has adequately addressed the purpose and audience as you understand them. Feel free to make lists and expand on arguments as you require for the notes to be useful to you. Although these notes are technically for your use as you build ideas for a report for Jody, it is possible that Jody may ask you for the notes and even show them to Capitani.

## Organizing Information

Using your notes, write up a detailed outline of the points you plan to make with Jody on Monday. Be sure to provide specific references and examples in the outline.

## Major Task 1

Prepare a brief, informal oral presentation on your findings for Jody. The presentation should outline your opinions about the manual and provide examples from the text itself. As an added precaution, it will be helpful for you to provide a formal summary in case Jody wants to use it as a quick reference as she talks with Capitani on Monday. Keep in mind what an introduction ideally needs to do for reader-users. Think through, as well, all of the contextual issues surrounding what you write and how you write it. Although Jody clearly wants your input, she has admitted to having already formed her opinion of the manual. And though she says she doesn't want you to feel as though you have to agree with each other, she's obviously decided to ask you, at least partly because she's curious about how you'll respond. Obviously, your job does not rest on how well you perform this task; however, the intangibles (such as making certain kinds of impressions) are the things that often help to persuade people about your abilities. Given the contextual elements, think through what appropriate appeals you may make in a summary of your findings.

## Follow-up Task

Jody appreciated your careful, detailed examination of the manual. The two of you largely agreed about the effectiveness of the technical manual, and Jody told you later that she used your summary as a point of reference when she spoke with Cappy later on Monday.

A week later, Jody tells you that Cap-Hays decided to adopt Speed Check. You hear nothing for two weeks. Then Jody comes to your office. Capitani was so impressed with your thoroughness that he has asked if you would have the time and interest in accepting a freelance job—on your own time—to revise the Speed Check manual. While the program is important to Cap-Hays as a labor management coordination system, the manual clearly has some problems that

Capitani would like to work out before they implement the program in six weeks. Jody says that there is no conflict of interest if you accept the offer, and the decision is completely up to you. However, she does warn that this could really eat up your free time.

Capitani has asked that you write him a letter accepting or declining. If you accept, he would like you to propose a salary for your freelance work as well as a timetable for when he can expect the work to be completed. Right now the manual has 84 pages, all of which contain about as much text as you saw in the introduction. You must consider whether you wish to accept the freelance work or decline it and write a letter to Capitani explaining your choice.

Capitani's address is:

Joseph Capitani, CEO
Cap-Hays Manufacturing
P.O. Box 22164
Ginger, Alabama 01285

## Major Task 2

Using the introductory section of the manual do a mock revision for Capitani to illustrate the types of changes he may be expecting for the document. Include a memo that briefly explains the nature of these changes. The challenge in writing the memo is to remember that Capitani has little interest in design or rhetorical theory. He simply wants a document that speaks to his workers; however, he is also the kind of man who wants to have an active role in all changes in his company operations. As a result, Capitani will want to be able to speak intelligently about whatever changes are made, if necessary.

## Case Appendix: Speed Check Introductory Pages

# INTRODUCTION

**OBJECTIVES:**

- **Learn how to use your terminal and how to log off and on Speed Check**
- **Learn about Speed Check's many uses**
- **Explain how commands and menus work**

**Welcome**

Welcome to the Speed Check training class! You are about to become one of the elite members of a fast-growing group of people who know and use Speed Check to their advantage to eliminate problems of collecting attendance and labor information. We hope you'll enjoy the next few hours as you learn about our system. We believe that with the assistance of Speed Check you will be able to better manage your area and your employees' labor record. It does require your commitment to the outlined procedures, and with that commitment you will find Speed Check to be a tremendous tool for you.

**You will have instant access to:**

- Who is at work
- Why people are absent
- Where people are working
- Hourly wages and deductions
- Calculate/transfer hours into payroll

**The objectives of Speed Check are to provide:**

- A lot of useful information about employees
- That is accurate
- And current
- With almost no paperwork

**THE TIME AND ATTENDANCE MONITOR**

* Used by employees to clock in and out and clock in and out of a division
* Replaces timeclocks and hand written and hand-figured payrolls
* Supervisors know who is at work and who isn't
* Tracks vacation, sick, and personal hours
* Basis for effective cost accounting
* Can be recalled for historical accuracy

**Payroll Monitor**

* Transfers attendance information to the payroll system
* Eliminates duplication and mistakes
* Eliminates manual payroll sheets
* Saves time
* Can be recalled for historical accuracy

**Speed Check Main Commands**

All commands start with the first two letters of the action:

| | |
|---|---|
| **-AD** | **ADD** |
| **-CH** | **CHANGE** |
| **-LI** | **LIST** |
| **-UT** | **UTILITY** |
| **-DE** | **DELETE** |
| **-RE** | **REPORT** |
| **-TR** | **TRACK** |
| **-RC** | **RECALL** |
| **-BA** | **BACK ONE SCREEN** |

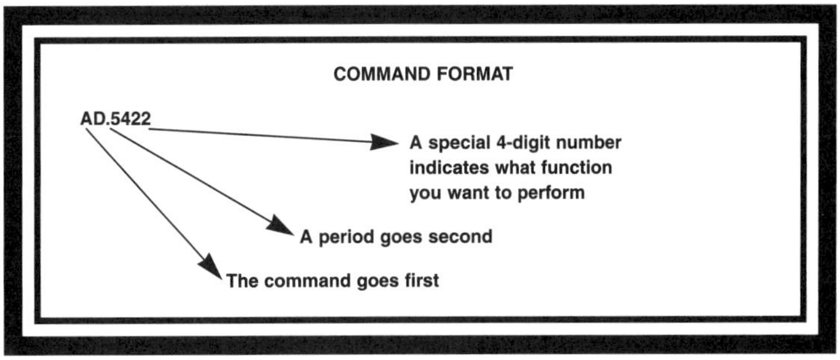
COMMAND FORMAT
AD.5422
A special 4-digit number indicates what function you want to perform
A period goes second
The command goes first

**Logging on for the first time**
You will be given instructions on how to log into the Speed Check system in your training session. Please write them here:

***SYSTEM LOGON PROCEDURE:***

In order to log onto the system you must be at a dollar prompt ($) to begin typing our open command.

Type the Open Command as follows:

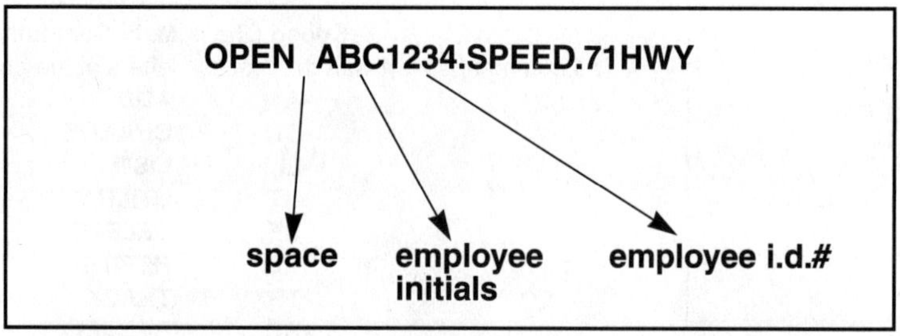


The system will then ask you for your password. *You will enter this password each time you log on, so do NOT forget the six-digit word you choose.*

**Your Terminal**
Listed below are some general rules to follow as you use your terminal:

1) Speed Check only works with UPPERCASE letters. Always make sure that the CAPS LOCK key is depressed for typing words.

2) You can correct errors by typing the BACKSPACE or DELETE key.

3) There are four arrow keys on the keyboard. Don't use them.

4) When you are done typing, hit the carriage RETURN key. The computer will not respond to the command or record any data unless you have hit RETURN.

5) Do NOT use the TAB key.

6) Do NOT use any keys marked INSERT LINE, INSERT CHARACTER, etc.

**Introductory Workshop**

In this workshop you will practice what you have learned from this section.

**Step 1:** Make sure the power to the terminal is on.

**Step 2:** Using the logon instructions, please log into the system.

**Step 3:** You will see a welcome message and then a message to hit the RETURN key. Do so now.

**Step 4:** You are now at the main menu. After the $ prompt, SELECT A FUNCTION KEY, type **LI888.** What happened? Why? Write your answer here.

________________________________________

________________________________________

________________________________________

**Step 6:** Follow the instruction on the screen until you are back at the main menu. Now type **LI.008.** This is the command to list an employee's detail information. You will now see a prompt to ENTER YOUR CLOCK NUMBER. Please enter your employee number. You will see another prompt for EMPLOYEE NUMBER. Please enter **0001**. You should see some information about employee 0001 on your screen.

What is employee 0001's name? ______________

Which shift does employee 0001 work? __________

What department is he or she in?______________

When was his or her last sick day? ____________

**Step 7:** You should now be at the EMPLOYEE NUMBER prompt again. Do you remember how to get out of this command when you are finished? What do you type to go back to the menu? Please type that now.

**Step 8:** At the SELECT A FUNCTION KEY prompt, please hit F3. What happened?

________________________________

**Step 9:** Use the function keys to go back to the main menu.

**You have just completed the introductory training session for Speed Check. Congratulations! You are ready to move on to the next section.**

☺

# Summary of Evaluative Criteria for Major Tasks in Case 1

| | **1 *Unacceptable***<br>Insufficient answer to assignment expectations | **2 *Below Average***<br>Inappropriate or ineffective verbal/ visual choices limit document success | **3 *Meets Task Expectations***<br>Has answered objectives of assignment, but individual components could be strengthened | **4 *Above Average***<br>Few flaws, document meets expectations, but could benefit from more attention to detail | **5 *Excellent/ Professional***<br>Few or no flaws, demonstrates keen insight into case subtleties and details |
|---|---|---|---|---|---|
| **Purpose/Key Points**<br>▪ Identifies and defines purpose (explicitly or implicitly)<br>▪ Articulates key points clearly<br>▪ Demonstrates an ability to analyze text extracted from context<br>▪ Works easily from notes in oral presentation | | | | | |
| **Context**<br>▪ Identifies/defines context and situational constraints<br>▪ Demonstrates awareness of document situatedness | | | | | |
| **Audience**<br>▪ Identifies/defines audience<br>▪ Establishes appropriate tone<br>▪ Understands technical details enough to communicate effectively | | | | | |
| **Organization**<br>▪ Demonstrates analytical insights<br>▪ Develops a clear line of argument<br>▪ Employs identifiable, appropriate pattern of organization | | | | | |
| **Design**<br>▪ Demonstrates awareness of visual design issues in analysis | | | | | |

CASE
2

# *Hiring New Colleagues: Aviation International's* Training Challenges*

This case asks you to assume a leadership role for technical communication trainees at Aviation International, a large commercial airline. You must analyze an "assignment" given to prospective trainees during the hiring process for the company. You will analyze a logo and a memo that communicates technical information.

## Background

Aviation International (AI) is a large commercial airline that offers passenger and cargo flights to most major cities in the United States as well as to 86 major cities across the world. The main operations office for AI is centrally located in Chicago. Most management offices in internal and external communications, engineering, business, public relations, and commercial operations are located here. In addition, many of the orientation and training programs occur at the Chicago office; however, AI also has numerous training programs in flight assistance and commercial operations in other cities such as Phoenix, San Francisco, Boston, Washington, DC, New York City, and Orlando.

Note the initiatives that have distinguished Aviation International as a strong and efficient airline.

For over three decades, AI has maintained a strong reputation for efficiency and safety. The FAA has consistently given AI high marks for quality control and maintenance. While some of AI's planes are relatively old (AI regularly runs five DC-10 planes that are 20 years or older), they are in good condition and have been grounded only in common situations facing most aircraft when there was question about a mechanical problem or adverse weather conditions. Over the years, AI has cultivated a strong business and frequent-flier clientele. AI was one of the first major airlines to create what has become the "business class" for traveling business people and to use the class in its overseas flights. In addition, in 1981, AI invested almost $350,000 to train all chief attendants in emergency medical response. AI was the first commercial airline to boast professional health care benefits to each flight. In 1989, AI offered a special service

* The author wishes to thank Darrell P. Chiavetta at Rockwell International for his help on the technical components of the case, and Tim Seydel at Buena Vista University for his graphic work.

for the critically ill that enhanced the emergency medical training bonus in its chief flight attendants. Those who were critically ill and seeking medical treatment in another state could fly free in a specially designed private section of the jets. One family member or attendant could fly half price. AI was heralded as "wings of mercy" in *Business Week* for the move.

Consider what you've read and observed about problems facing many different commercial airlines in the U.S. Are AI's problems comparable?

Recently, however, AI has suffered a few public relations problems. First, flight attendants for AI participated in an eight-week strike in a dispute over wages, benefits, and training opportunities. AI employees were not the only flight attendants who participated in the nationwide strike, but union organizers made very public statements about AI's uneven medical training for flight attendants and its less-than-adequate retirement packages. Strikers actively blocked replacement flight attendants from crossing picket lines, and several arrests were made in both Phoenix and New York. Though the strike was resolved relatively quickly, it was not without its cost to AI.

In March of last year, a near disaster was barely averted when faulty wiring caused a compressor blade to expand and break on a 747 bound for Dallas–Ft. Worth from Los Angeles. The jet lost the two left engines and was forced to make an unscheduled landing in the desert 70 miles northeast of Phoenix. While the landing was a little rough, there were no injuries reported among the 361 passengers aboard. Later, the pilots and crew for flight 1033 were praised by the press for their bravery and quick response, but the FAA's investigation into the problems that led to the emergency landing determined that the wiring had been improperly checked prior to takeoff. For the first time in its history, AI felt the pressure of public scrutiny over its safety standards.

Finally, in the most recent annual report issued last month, AI profits were reported down nearly 9 percent from last year at this time. Increases in costs coupled with loss of revenue resulting from airfare wars and increased low-budget competition have contributed to AI's dive in profits. The AI board of directors is now considering cutbacks in middle-management personnel though no decision has been reached at this point.

## The Situation and Your Role

Your role is defined here. What do you expect this position requires of you?

Although AI is clearly experiencing some changes, most of the operations are continuing as normal. You are one of the senior communication specialists for AI; you have been with the company for eight years. Your job entails, among other duties, signing off on all high-level technical documents and public statements, as well as supervising revision schedules for manuals and public documents. In addition, you help with the hiring and training of new communication specialists at the Chicago home office.

Communication specialists for AI are in charge of designing and writing manuals, brochures, and instructions for everyone from frequent-flier customers to airstrip signal personnel and from flight attendants to mechanics. In addition, the Chicago communication office team is in charge of publishing the organization's newsletter/magazine. Historically, AI has hired outside consultants to produce the high-caliber glossy documents largely for the public (including the in-flight *Aviation International Eagle* magazine) and to design the company's identification package (including the logo). Because of financial constraints, however, the AI board of directors has suggested that more of the high-profile technical and organizational communication tasks be completed in-house. Thus, because of your senior position, you and your colleague, Abigail Bennett, are faced with hiring two entry-level communication specialists who can help to meet AI's technical and professional communication needs both internally and externally.

You and Abigail Bennett came to AI at approximately the same time and are long-time collaborators. You genuinely respect Abigail's insights into AI initiatives. So, when you meet to discuss the hiring process, you ask her, "If AI needs more major projects done in-house, shouldn't we be advertising for something other than entry level?"

Abigail sighs as she hefts a stack of résumés and portfolios onto the table between you. "Well, I think the assumption is that the senior communication specialists will have more of a hand in the stuff that's normally contracted out. The people we hire probably need to be groomed to do the tougher design stuff, but they won't start out with it."

This makes sense, though you're a little concerned with the ramifications this change of policy has on your own positions. "If we're doing more of AI's publications in house, how are we going to be expected to fulfill the client counseling, personnel training, and other work that we do? I'm just not sure what they're thinking here."

Abigail proposes a change in hiring procedures and initiates the collaborative process here.

"They're thinking this is going to save them money, which it will likely do, if you consider it for a few moments. We may be doing more work initially, but if we hire right, we should be able to groom the new folks to pick up the slack where necessary. Keep in mind this is two positions we didn't have before." Abigail pauses as she picks up a brightly colored portfolio. "You know," she adds, "I've been thinking that we ought to maybe add a dimension to the hiring process this time around. I mean, if we're really looking for initiative, creativity, and a clear sense of the organization's communication needs, then besides the standard education and experience criteria we normally use shouldn't we have a component to this process that helps us assess initiative and creativity?"

"Of course that makes sense," you respond. "Should we give them some sort of assignment? I once got a copywriting job because I could copyedit what they gave me in the interview in five minutes."

Abigail nods enthusiastically. "I like that. An assignment. Hm..." she pauses and looks at the portfolio in her hands. "What about asking those we interview to create a new company logo, put it on letterhead, or write out a memo to a department head regarding some sort of procedural change? If we gave them the assignment before they came to the interview, they could bring it with them and present what they came up with as well as their rationale for the design."

"A new logo and rationale would certainly help to see how the interviewee understands the nature of Aviation International," you agree. "But I think it's more important to get them to actually communicate something technical—possibly what you suggest, a procedural change—because I doubt we'd end up using some sort of logo designed by a new hire when we have paid thousands for consultants to do it before."

Abigail nods. "You're right. It's unlikely we'd use it, but who knows? What if something came in that was just the ticket? Talk about saving AI some serious money."

The entry-level position you are hiring for is more clearly defined here. Note the skills that you and Abigail are privileging.

"But the job won't ask them to do that," you argue. "These people will be hired to communicate technical information clearly and succinctly. The memo is probably more important."

"I see the skills as equal," Abigail persists. "If we are indeed grooming these new hires for the more complex in-house jobs, they need to be able to design an annual report for stockholders as well as they can edit an altimeter repair manual. They also need to be able to design, write, and give good presentations. We're out of the specialist-who-just-does-one-thing era."

You hold up your hands in defense and laugh. "OK, OK. I have no problem with the logo part, but I can tell you that I think I'll end up privileging how well they communicate the technical information. I imagine we'll balance each other out. We always do."

Whenever you consider passing a judgment, it is essential to clearly define the criteria by which you will evaluate your subject.

Abigail nods and glances again at the portfolio in her hands. "You know, we probably ought to think about setting up some criteria for judgment, just to make sure we're on the same page." She takes out her yellow legal pad and writes: *Criteria for Sample Document* with a line underneath for emphasis. "OK, what next? Well, obviously, they need to have the information right. Then again, we'll likely give them most of that in the prompt." She shrugs and writes *Content (accuracy).*

"I think it's important to look at how they've judged their audience. Put down *audience appropriateness.*" Abigail writes. "The memo is going to need to be pretty concise, but well organized, too. Half the people who might read this will get through the first couple of lines, and if they don't see what the memo's for, they'll assume it isn't really directed at them."

Abigail looks up and nods, "Like that back check order for the exterior." Two years ago, all flight technicians had received a special memo requiring a back check of each jet's exterior before takeoff. However, technicians had complained that the memo was too long and difficult to understand because it was laden with superfluous information and rationale. As a result, many of the technicians chose to ignore the memo. Eventually, after flights were delayed and technicians explained the situation to superiors, procedural directives were scaled back considerably; communication specialists were instructed to privilege brevity and accuracy.

Abigail made more notes on her pad. "I think design's important."

"OK, you can be the design guru, but I'm the stickler for mechanics and professionalism. The document needs to be clean and accessible. They need to demonstrate that they understand AI and that they are taking the opportunity to work for AI seriously. I also want to make sure they understand the terms they're using if we throw any technical language at them."

Abigail continues to write. When she is finished, your collaboratively created list of criteria looks like this:

*Criteria for Sample Document*

*What we want:*

- ✔ *Attention to content (accuracy)*
- ✔ *Awareness of audience (appropriateness of tone, style, length, etc.*
- ✔ *Organization (clear movement/purpose)*
- ✔ *Attention to design*
- ✔ *Professionalism (understanding of terminology, mechanical correctness, technical detail)*

As you finish your meeting, you both agree to sift through the résumés and portfolios and rank-order your preferences for interviews. Once you and Abigail have decided on five people to bring to Chicago for an interview, you will send out the assignment.

The assignment is one you have created based in part on Flight 1033's mishap last spring. Because one of the FAA's criticisms of AI hinged on the overall lack of communication between technicians and flight crew, AI committed to strengthening relations across professional lines and scheduled a series of monthly informational meetings that were incorporated into the retraining and continuing education program. These all-day meetings were designed as conference gatherings at each major AI office and included technicians, flight crew members, management, and engineers. At each meeting, a variety of teams participated in discussion panels and small-group presentations on important issues, new techniques, and refresher skill building. Communication specialists created short packets of summaries from the meetings that could be distributed to employees across the country who could not attend. AI management agreed to continue the meetings for one year and then reevaluate the company's needs.

One of the first meetings dealt with the 1033 flight problems and was led by a team that included Flight 1033's copilot, an engineer, and a ground crew technician. The copilot, Scott Armstrong, discussed the importance of applying Bernoulli's law to the techniques he used to help land the jet safely. According to Armstrong, when the left engines failed, the pilot then used the rudder to slow and speed for an asymmetric thrust to employ Bernoulli's law. Bernoulli's law explains the relationship between pressure and velocity of moving fluids; however, the law also directly explains the lift that permits airplanes to fly. When air travels across the wing (the upper surface of which is more curved than the lower), pressure on the upper surface is reduced as the air travels faster. Pressure subsequently increases on the lower surface and provides the lift for the aircraft. The copilot explained that basic knowledge of Bernoulli's law was important to determining compensation given the change in pressure and speed that occurred when they lost the two engines.

The assignment for the job candidates is outlined. To better understand Bernoulli's law, do some background research on the physics involved.

The assignment that you and Abigail agree upon asks the interviewee to research and briefly summarize Bernoulli's law in a short memo to the senior communication specialist, who will include the information in one of the post-meeting packets for national distribution. The writer should understand that management, pilots (both experienced and relatively new), engineers, and technicians will all have access to and possibly read the summaries included in the post-meeting packet. Although most of the technical and skilled AI employees will have heard of Bernoulli's law (and many will have direct experience applying it), there will be others in the audience who do not know or have technical understanding of it.

## Background Development

Consider the following questions as a way to evaluate what you've read in the case. Answer the questions on paper, or work in a group and discuss possible answers. You may use your responses to help you in subsequent tasks.

- Look again at the collaboratively developed list of criteria for the sample document. While the criteria are definitely important, at this point there are no parameters for judgment designed to help the readers respond to the samples created by the applicants. As a professional in the field for over eight years, you are unlikely, in fact, to need to outline what you see as clear organization or solid design skills. However, assume you must hand the screening job over to Abigail alone because you must be out of town. Abigail wants your input, and while she's worked with you for a long time, it would be helpful to her if she had an outline of how you would judge the criteria; that is, what would make a given application package "successful" or "unsuccessful"? Write out a brief list of your expectations for how the criteria will be best met. You may refer to your text for help in exploring the criteria.
- It would be helpful for you to fully understand Bernoulli's law in order to judge the accuracy of the applicants' samples. With a partner, scan your library for any other descriptions of Bernoulli's law. Summarize your findings so that you may refer to them as you review applicants' samples.
- Reread the background section describing AI. With the knowledge you have of AI's history, what kind of image do you believe the airline should communicate about itself? How would you describe its character? Write down your ideas and then exchange them with a classmate and compare what you both come up with. Are your ideas similar or dissimilar? How would these ideas inform how you read an applicant's design and text?

## Getting Started

You and Abigail have chosen the five candidates you would like to invite to Chicago for an interview. Over the phone, you have explained to the candidates that a packet of information about AI will be directly mailed to them. You have also explained the assignment you and Abigail have created. You and Abigail agreed, however, that an oral description of the task over the phone is not enough for the candidates to go on. You have agreed to write a cover letter to accompany the materials and briefly explain the task to the candidates. Using the first candidate's name, Joshua Marius, write a draft of this cover letter for Abigail to read and provide feedback. Marius's address is 2126 Admiralty Way, Washington, DC 20090.

## Major Task 1

Based on the criteria you and Abigail have determined, analyze the following memos and AI logos in the case appendix. How do the pieces address the needs of the airline, based on your understanding of its background? Write a brief memo to Abigail outlining your analysis. Remember that your perspective is important and that you are effectively working to persuade Abigail to see the candidates' work as you do.

## Major Task 2

Because the logos/letterheads and the memos were separated intentionally, you must evaluate each individually. You find that you prefer the design created by candidate 3 but the technical memo written by another candidate. How do you resolve this apparent conflict? Write an e-mail to Abigail explaining your preferences. Offer a solution to the problem.

## Major Task 3

Assume the role of a prospective interviewee. Based on the memos you see here, the assignment, and your understanding of AI, create your own letterhead and logo and write a brief memo that describes Bernoulli's law. Then exchange your work with a partner and assess each other's work based on the criteria outlined here.

# Case Appendix: Memos from Job Applicants

## Memo 1

AVIATION INTERNATIONAL

To: Abigail Bennett, senior communication specialist
From: Carter Simmons, communication specialist
Date: June 24, 1998
Subj: Summary of Bernoulli's principle

The following is a summary of Bernoulli's principle, to be reported in the next meeting's summary (print date 7-27-98, as per your request.

> **Bernoulli's Principle**—explains the hydraulic lift (upward force) which occurs as increased velocity creates a decrease in pressure on the lower portion of the wing. The force allows the jet to lift off. Bernoulli's principle is a basic flight law and should be understood by all pilots and engineers.

Please contact me if I may expand this summary or answer any questions.

## Memo 2

AVIATION
INTERNATIONAL

**Memorandum**

Date: June 26, 1998
To: All Aviation International Senior Communication Specialists
From: Joshua Marius, communication specialist
Re: Bernoulli's Law Summary

Named for the famous Dutch-born Swiss mathematician and physicist Daniel Bernoulli, Bernoulli's law refers to the mechanics of velocity and pressure specifically as they explain aerodynamic lift. The theory, originally applied to the physics of moving fluids and gases, correlates speed of movement with pressure. According to the *Academic American Encyclopedia,* Bernoulli offered the example of water flowing through a pipe. "If water... flows through a horizontal pipe of varying cross section, the water must flow faster in the narrower regions. The pressure must be greater in the wider regions, because the walls of the pipe must exert a force to accelerate the water on its way to the constriction" (223).

Applied to flight, then, Bernoulli's law argues that because the upper surface of a jet's wing is curved, air pressure necessarily decreases as velocity increases. The increased pressure on the underside of the wing forces the jet to lift once it has hit the appropriate velocity.

# Summary of Evaluative Criteria for Major Tasks in Case 2

| | *1 Unacceptable*<br>Insufficient answer to assignment expectations | *2 Below Average*<br>Inappropriate or ineffective verbal/ visual choices limit document success | *3 Meets Task Expectations*<br>Has answered objectives of assign-ment, but individual components could be strengthened | *4 Above Average*<br>Few flaws, document meets expectations, but could benefit from more attention to detail | *5 Excellent/ Professional*<br>Few or no flaws, demonstrates keen insight into case subtleties and details |
|---|---|---|---|---|---|
| **Purpose/Key Points**<br>▪ Identifies and meets purpose<br>▪ Articulates key points clearly<br>▪ Demonstrates care-ful analysis of Aviation Intl. and its public image<br>▪ Demonstrates appropriate use of technical language skills and design elements | | | | | |
| **Context**<br>▪ Identifies/defines situational con-straints<br>▪ Demonstrates com-munication skills in collaborative part-nership | | | | | |
| **Audience**<br>▪ Identifies/defines audience and meets identifiable needs<br>▪ Establishes appro-priate tone<br>▪ Reads and responds to others' work appropriately<br>▪ Correspondence offers ample infor-mation to partner | | | | | |
| **Design**<br>▪ Demonstrates awareness of visual design elements of task<br>▪ Demonstrates an awareness of design options & technological aids in the development of these options | | | | | |

# CASE 3 Designing a Plan of Action for the St. James Downtown Development Process

St. James is a small northeastern town whose leaders want to revitalize its downtown business district. The St. James City Council has hired a consulting firm to analyze the town's needs and offer possible revitalization projects. As one of the consultants, you are asked to create a tentative communication plan of action chart.

## Background

While St. James is a fictitious city, you may certainly still gain much insight into the general area by looking at a map or an atlas that gives you a picture of this region. It is useful to have some insight into upstate New York's natural resources

Note that a turning point in the makeup of this community is outlined here.

St. James is a town of approximately 25,000 located in upstate New York in the relatively rural area of the Finger Lakes region. Until the 1970s, the town relied upon three major manufacturing companies for the bulk of its revenue and employment—Orinoco Paper, a large-scale paper mill; a Ford auto plant; and Wolverine Technologies, a mid-sized electronics manufacturer.

In the early 1970s, however, the small town was hit hard by the nation's economic woes, and Orinoco Paper laid off nearly half of its 670 employees. In addition, Wolverine Technologies officially closed its doors after several years of struggling (and failing) to maintain contracts with large distribution companies such as Sony and Hitachi. The loss of Wolverine Technologies spelled near defeat for the struggling community, and St. James saw a corresponding increase in crime and dropout rates.

Things began to turn around for the town, however, in the early 1980s, when investor and long-time resident Chandler Moss challenged the city to take advantage of its natural resources and tourism potential. With the nearby Appalachian mountain range and densely wooded surrounding land, the city's beauty was undeniable A task force, designed by the city council and headed by Moss, initiated a nationwide marketing campaign for the community to bolster tourism. In addition, the city sought and won several grants for community improvement. Most of the money from the grants went to aesthetic improvement of the city's parks and trails. Moss himself offered an unprecedented $2 million matching grant to the St. James Chamber of Commerce. The city then initiated an aggressive capital campaign among current and former residents and was successful in surpassing the $2 million goal by 24 percent. With the money, city officials and Chamber of Commerce members established a low-interest loan fund for entrepreneurs interested in starting up small businesses. Those proposed businesses that could show a connection with tourism were of highest priority. The city's Main Street (the heart of its downtown) saw several new novelty and specialty shops specifically geared to tourist business move in—a bicycle and sports apparel store, an antique store, an art gallery and print shop, and an import wine and cheese shop that offered a small coffee and dessert outdoor café.

In 1987, the Wolverine Technologies factory was purchased by a Detroit-based electrical manufacturer, and the plant, now called Fortmann Electronics, was resurrected. While the city could not recoup all of the jobs lost in the Wolverine shutdown, Fortmann Electronics offered 275 new jobs to the community. By the end of the decade, St. James's economy was growing and the city was decidedly healthier than it had been just 10 years before.

How would you describe the conflict here? Who/what is at odds?

In 1991, a strip mall with 16 specialty shops and two base stores opened at the far north edge of town. The two base stores (located at both ends of the row of shops) included a chain grocery store and a large discount department store. The latter offered merchandise such as clothing, sports equipment, toys, electronics, lawn and garden materials, cooking utensils, and dinnerware at significantly reduced prices. In addition, the discount store offered services such as a pharmacy, custom tailoring, and domestic auto repair. A significant number of the businesses in the downtown district offered either similar products or services as those in the strip mall or the discount department store. But since the strip mall was strategically located for convenience next to a grocery store, and since the prices at these locations were generally lower than those at businesses downtown, the Main Street shops began to lose business. Because the downtown businesses had only recently been reinvigorated because of the efforts from Moss and the Chamber of Commerce, city officials began to worry about the impact of the strip mall on other interests such as tourism. At a city council meeting in May 1994, one council member was quoted in the local paper arguing, "People don't go on vacation to shop at the mall. They can do that at home. They get away to experience the culture and history of another place, and our downtown businesses should be designed to introduce tourists to the history of St. James as well as offer a unique shopping experience."

Several of the downtown business owners concurred and openly criticized the city council for "shooting St. James in the foot" by changing the zoning law that allowed the strip mall to be built. While others argued that the strip mall and discount store had provided important jobs and economic growth for the community, most agreed by 1996 that it was time to examine what measures might be taken to reinvigorate the downtown.

## The Problem

You work with a consulting firm called Wilde & Parks Associates, based in Boston, that specializes in city redevelopment projects. Wilde & Parks generally sends in a team of three to five consultants for a given project (depending on

Note that Wilde & Parks's consulting process is outlined briefly here. If you were to create a flowchart with the steps Wilde & Parks goes through in evaluating and responding to a client's needs, what might it look like, based on what you read here?

size and scope) to examine and evaluate a variety of components to the project (e.g., landscaping, in-fill, architecture/redevelopment, traffic, population demographics, and economic potential, to name just a few). The consulting firm's chief obligation to clients is to examine the perceived needs of the community alongside the actual assets and limitations of the area. After a careful evaluation, the consulting firm then proposes a specific course of action for the clients. If the clients accept the proposal, Wilde & Parks lobbies to manage the implementation of the plan. Wilde & Parks have never had a plan accepted without also subsequently being hired for the implementation stage.

Note that your role is defined here. What does it mean to you that you have a background in both economics and communication?

The consultants for Wilde & Parks Associates vary in background—some have architecture degrees, others are former city planners. Some are communication specialists, others are engineers or landscape artists. Your background is in both economics and communication, and you have frequently served as a team leader for various projects. Your dual discipline background makes you a good spokesperson for consulting teams because you understand the subtleties of economic planning and you can articulate these realities to laypeople and professionals alike.

The city of St. James sent out a call for proposals among prospective consultants in the spring of 1997. According to the request for proposals letter, drafted by the city council, the city of St. James sought a consulting firm with ample experience with community populations between 15,000 and 40,000. In addition, city officials emphasized the importance of incorporating a *participatory* planning process that would encourage a wide variety of community members to be actively involved in the discussion about the future for downtown.

Consider talking with a professional with some interest and/or expertise in city planning or management. Develop some questions for that person about trends toward more participatory planning in city development initiatives. What are the benefits and drawbacks of such approaches?

When Wilde & Parks received the request for proposals, the vice president, Cameron Wilde, requested that you lead the team and draft the proposal. You took the initiative to call the St. James City Manager, Art Reeves, to ask a little more about the community and request that a package of information be forwarded to Wilde & Parks.

After some perfunctory discussion about deadlines and requirements, you asked Reeves, "I notice your interest in a participatory planning process. What's the city's purpose in this?"

"Well," Reeves replied, "the city offices and the city council have been pretty heavily criticized over the past few years for certain decisions. We decided we wanted some involvement outside those ranks so folks might begin to understand what goes into making change in a place like St. James."

After the conversation and a brief trip to St. James, you meet with your team and in three weeks you have drafted a proposal for the consulting job that outlines a year-long participatory study. Your team is invited to St. James in July to offer a presentation based on the proposal that outlines the study you have envisioned for the community. Based on the quality of the proposal and its emphasis on community participation, you are offered the consulting job in August and you begin planning your first meeting for the following month.

Your team consists of yourself and two other associates—Jean Bral, a landscape artist, and Jeremy Kurtz, a city planning expert. You have also asked that Keaton Carver, an architect with Wilde & Parks, help the team out with drafting later in the process. You have worked closely with Bral in the past and respect her professionalism and analytical abilities. Sometimes her skill with an audience is also superior to yours, so you enjoy having her expertise to rely on, especially in processes you know will involve participation from a variety of community members. Kurtz is new to Wilde & Parks and this is his first team-consulting experience. You know Carver well, and his drafting work is among the best at the firm. Generally, you have a great deal of confidence in the people who will work with you on the St. James project.

Note how the initial purposes for your team are briefly defined here. How are lists useful? Limiting?

Your first order of business with your team is to develop two lists that will serve as the basis for your planning process over the next year. The first list consists of the primary individuals and groups of people to whom you must answer throughout the process and with whom you must communicate findings. The second list is a rough draft of the steps and benchmarks you will use to indicate forward movement in the planning stage of the project. While both of these lists are usually drafted by the team leader after some initial meetings with city officials, you believe this is a good opportunity for Kurtz to get his feet wet with this project. Thus, you take him with you to your first meeting with the St. James city officials and city council members. Chandler Moss also sits in on the meeting.

After introductions, you explain to those at the meeting the importance of reaching as many people as possible to ensure investment in any sort of redevelopment project in the future. But the key to successful communication in this situation is not just "getting the word out," but involving people appropriately by developing their ideas and sharing information at appropriate levels.

"So what you're saying is that you can't involve all of the people all of the time, right?" Moss asks.

You nod. "Because a planning project of this magnitude takes time and sustained interest, we have to make sure that the pertinent information and ideas are channeled to the most appropriate audiences. For example, while the city-wide group of participants we hope to attract at our series of town meetings will certainly have an interest in the economic viability of various ideas, it's unlikely that they will want much more than the bottom-line sorts of figures. The city council, on the other hand, will require a more careful breakdown and analysis of the costs for each proposed change and project, in part because you're the ones who have to allocate the funds."

Kurtz jumps in at this point and explains, "What we're looking for here is a list of possible stakeholders groups, people who, either individually or in small groups, ultimately have a stake in the planning process and who we will need to communicate with in one way or another."

Are you given the impression that this list contains any sort of implied hierarchy of groups? Would you read the list differently if you knew there were? As an outsider, is it your responsibility to understand the political makeup of a community if you commit to working for it?

The group agrees to brainstorm over lunch, and you reconvene later in the afternoon. The city council provides you with the list (on the following page) and key contact people.

After the city council members explain each of the groups and offer general insights into each, you and Kurtz meet privately back in the hotel room to discuss the list.

"Do you think you'd add anyone here? Do you see any gaps?" you ask him.

"Well, I think that you can't exclude the city newspaper, and I'm not sure where they fit in the list, " he notes.

## Prospective Stakeholders List

**Downtown business owners**
Paul Breuch, Representative
Owner, Rags to Riches Antique Shop
President, Downtown Business Association
555-1452

**Downtown service representatives**
Shari Stucker, Representative
Women and Children Shelter Services
555-2222

**Downtown residents**
Autumn Pritchard, Representative
Neighborhood Coalition President
555-9843

**St. James community participants**
Leaders to be determined at town meetings

**City council**
Kevin Schleuter, Representative
City Council Chair, Lawyer
555-9169

**Chamber of Commerce**
Clare Kolat, Representative
President, St. James Chamber of Commerce
Owner, His and Hers Fashions (Forbes Mall)
555-7329

**Developers**
Chandler Moss, Representative
555-6682

You nod. "The newspaper will offer us a strong connection to the community, if nothing else, simply because we'll use it to advertise the progress." You add the *St. James Herald-Register* to your list. "Anybody else?" you ask.

Kurtz frowns as he concentrates on the list. "Reeves and his staff aren't included here, which sort of surprises me. The city offices are important and different constituents from the city council. I'd definitely include them. Seems as if they're left off."

Kurtz has detected what you saw as a potential problem immediately. It appears as though the city manager's office is almost deliberately left out of the loop, and you wonder if that doesn't suggest a political rift between the council and the city manager's office. You include Reeves and the city manager's office on your list. You and Kurtz then return home to Boston to confer with Jean Bral on the time frame.

Based on your examination of the city's goals for the planning project, your team has determined the following list of events that may serve as benchmarks for the process.

| Meeting Type | Frequency |
|---|---|
| **Town meetings**—These public meetings will be used as opportunities for community members to come together to identify strengths and weaknesses of the downtown, brainstorm for changes, and problem-solve as a group. | No fewer than three and no more than five. Meetings should last approximately two hours each. |
| **Small-group stakeholders meetings**—These meetings should involve those individuals identified as key stakeholders in downtown planning. Meetings will be divided according to identity of group and participants will outline specific concerns for downtown pertaining to area (e.g., Chamber of Commerce concerns for economic development downtown). | No more than two meetings per group. Meetings should last one to two hours each. |
| **City officials and Zoning Commission**—These meetings will be designed to help acquaint team with city history and understand city potential and limitations. | No limit on time or number. |
| **City exploration meetings**—These "walking tours" of various areas in St. James will help team members and participants to understand the total picture of the community and the context into which the downtown figures. Walking tours should be led by a variety of stakeholders including (but not limited to) business owners, residents, developers, city council members, and the mayor. | The number of these walking tours will depend upon interest among participants. |
| **City council meetings**—The team needs to attend at least one regular city council meeting and would like to schedule an opportunity for at least two other meetings to present findings and questions. | Three meetings. Time to be determined by city council. |

In addition to the scheduled meetings, team members would schedule time to study city documents and previous plans as well as meet with individuals by request.

## Gathering Background Information

Compare the documents you and other classmates gather from city offices. What can you learn from these working papers?

City development projects such as the one initiated by St. James are common across the nation. Most cities with populations of 5,000 or more have city manager's offices and a city council that are designed to address (ideally, in a proactive manner) city growth and development issues. As a class, or in small groups, take a trip to your local city offices and request copies of recent planning and zoning reports. These are public documents and should be made available to the public. If the city has a recent planning process document, request this as well. Examine these reports and what issues they address. What sort of expertise does the reader need to have to understand the documents? Or are they written in lay terms for public consumption?

## Major Task 1

Using the two charts you just obtained, plan out a tentative schedule for the next year that will allow the team time to gather information, analyze data, and report findings to the appropriate people. Make this schedule as complete as possible using dates beginning with September 20, 1998, and ending at approximately the same time the following year. You should make this schedule as visually accessible and understandable as possible, particularly since you will use it first to explain to your fellow team members how you have envisioned the project. Be prepared to explain your choices orally.

## Follow-up Task

Outline your schedule to Kurtz and Bral in an informal oral presentation, explaining your choices and illustrating what you see as the possible benchmarks for the process. Benchmarks are the tangible outcomes you might produce at any given point in the process (e.g., some sort of statistical analysis or a report—internal or external).

## Major Task 2

Create a separate chart that adds in the possible opportunities for communication within the process you have outlined. That is, after a public meeting, for example, would you have the opportunity to write up a press release for the newspaper explaining to those who couldn't attend what was accomplished? Or after several walking tours, might it be logical to write up a report to the city manager (or perhaps to be used in a final report to the city council) detailing what you learned about the area? Indicate not only what type of communication you foresee for this point in the process (e.g., a technical report, a news article, an oral presentation, a proposal), but who you see to be primary and secondary audiences for what you produce. Feel free to be as detailed as you can be, but acknowledge that nothing you create here is set in stone. It should be flexible according to the process itself. Create a chart that reflects the outcomes or written/oral forms of communication you anticipate. You may assume that this chart will be used among the team members and may also be presented in the first town meeting as a possible glimpse at the process based on what you know now.

## Follow-up Task

Discuss with your classmates possible problems you could experience by making public either of the two documents produced in Major Tasks 1 and 2. How are these documents inherently political? What issues/people/areas might they look as though they privilege?

# Summary of Evaluative Criteria for Major Tasks in Case 3

| | *1 Unacceptable*<br>Insufficient answer to assignment expectations | *2 Below Average*<br>Inappropriate or ineffective verbal/ visual choices limit document success | *3 Meets Task Expectations*<br>Has answered objectives of assignment, but individual components could be strengthened | *4 Above Average*<br>Few flaws, document meets expectations, but could benefit from more attention to detail | *5 Excellent/ Professional*<br>Few or no flaws, demonstrates keen insight into case subtleties and details |
|---|---|---|---|---|---|
| **Purpose/Key Points**<br>▪ Identifies and meets purpose<br>▪ Articulates key points clearly<br>▪ Demonstrates careful analysis of St. James's situation and Wilde & Parks's approach<br>▪ Demonstrates appropriate use of verbal and visual elements for major tasks | | | | | |
| **Context**<br>▪ Acknowledges importance of history<br>▪ Initiates research for outside information | | | | | |
| **Audience**<br>▪ Identifies/defines audience(s) and meets needs<br>▪ Establishes appropriate tone<br>▪ Anticipation of outcomes or benchmarks is appropriate to process, goals, and identified audiences | | | | | |
| **Design**<br>▪ Demonstrates awareness of visual design elements of task<br>▪ Demonstrates an awareness of design options | | | | | |

# CASE 4

# *Communicating to a Variety of Audiences: Describing Zond Corporation's[1] Unique Reusable Energy Source*

Zond's generator run by wind power is being introduced to the small city of Indianola, Iowa as an alternative energy source. The Zond promoters need to communicate how the machine works as well as its benefits to the community, but quickly discover that those involved with the project have conflicting informational needs. Because community buy-in is important to the success of the alternative energy program, promoters must determine the most effective way to deal with varied audiences.

## Background

Zond Corporation was founded in 1980 by a pioneer in reusable energy sources, Jim Dehlsen. Dehlsen earned B.S. and M.B.A. degrees with honors from the University of Southern California. After founding a successful business and financial consulting firm, Dehlsen went on to found Triflon Company, Inc., which introduced a new lubrication technology based on the use of micronized Teflon particles. From there, Dehlsen became interested in the research and development of renewable energy sources; specifically, he began to investigate wind power electric generation. Together with Craig Anderson and Kenneth C. Karas, Dehlsen formed Zond Windsystems Operating Corporation, incorporated in 1982.

Why is it important to understand a corporation's history? Discuss with classmates.

Zond quickly became a leader in the windpower industry as it developed, constructed, and maintained scale windpower plants for itself, utilities, and independent power producers. Initially, the company, located in Tehachapi, California, commissioned and purchased high-power wind turbines designed and engineered by a Danish factory called Vestas. Eventually, Zond's own engineers perfected the turbine design to increase efficiency and began manufacturing wind turbines of their own.

[1] The author wishes to thank Jim Dehlsen, Zond's founder and chair, and Ken Hach, midwest regional manager for Zond Corporation. Information for the background section of this case on Zond Corporation was made possible with the help of Ken Hach and publications and videos created for the education and promotion of Zond Corporation. While the background sketch of Zond Corporation is based on factual information, the situation in Indianola, Iowa is hypothetical.

You can learn more about the initial projects in California by accessing Zond's World Wide Web site.

Zond's initial projects were based largely in southern California because the corporation worked directly with Southern California Edison, Co. and the California Energy Commission. The air over California's deserts and valleys made the state an appropriate launching pad for Zond's first large-scale projects. Because the air in southern California warms during the day and the heated air rises, subsequently drawing cooler marine air inland, the winds that are driven by this process channel through passes and flow over the lower hills. Winds naturally accelerate, then, as they flow over the lower hills in the coastal ranges. California's high mountain ridges and peaks are exposed to fast-moving currents and also have strong and persistent winds. Thus, California offered the appropriate wind currents with which to begin the large-scale projects for Zond. Since its initial "Victory Garden" project, however, Zond has expanded to a variety of states with favorable wind conditions like California's.

It may be helpful to do some initial research into all types of energy to have a sense of what the U.S. predominantly uses. Make your first source a recent world almanac, which should summarize types of energy and the extent to which they are used.

Zond's history is consistent with that of renewable energy sources in the United States over the past two decades. Renewable energy sources, and windpower specifically, have experienced both apathy and surges in public interest based largely on the international and domestic political climates. In the late 1970s and early 1980s, as this country experienced the oil crisis and tenuous political relations with foreign countries rich in oil, public support of renewable energy sources and new tax credits established by the Carter administration allowed Zond Corporation and other corporations like it to form and begin researching alternative energy possibilities. Through these early years, Zond discovered Vestas, which had already begun to perfect wind turbine designs that could harness energy efficiently while still withstanding the enormous forces of the weather to which the turbines were daily subjected. In 1985, Zond installed 1,100 machines in four different sites, using a "wind wall" design unique to the company.

Later that same year, however, with growing strength in OPEC, renewable energy tax credits were terminated and federal government subsidies were all but eliminated. Zond, like other companies connected with the renewable energy movement, suffered under federal cutbacks and began to consider seeking European support. In the mid-1980s, however, the nuclear plant at Chernobyl in the former Soviet Union experienced a disaster that negatively impacted attitudes about nuclear energy across the globe and reinvigorated the private campaign for alternatives to fossil fuels and environmentally harmful energy sources. Later, after the Persian Gulf War, investment in renewable energy became even more vital, and Zond Corporation was able to initiate the Sky River Project, which would provide enough clean energy for a residential population of 300,000 using high performance blades and Zond's custom design Z-class turbines. Estimates suggest that the wind energy produced by the Sky River Project alone replaced 12 million barrels of oil. In addition, one Z-40 turbine, over its life of approximately 25 years, will displace 27,000 tons of coal for a cleaner and more cost-effective supply of energy.

Through the 1990s, attention to alternative energy sources has expanded; Zond Corporation has followed suit, becoming the largest renewable energy company of its kind in the world.

## The Situation

For a visual introduction to the Z-class turbines, check out the photo images and graphics of the turbine design on the website for this case.

As Zond Corporation has continued to expand and interest in harnessing wind as a renewable energy source has increased, potential investors have sought out Zond as an intriguing alternative to other business ventures. In addition, Zond has expanded potential project sites as the Z-class turbines have become more efficient.

In 1994, an eleventh-grade honors science class at Indianola High School in Indianola, Iowa, began studying renewable energy sources. The science teacher, Jan Mills, encouraged her students to consider the sources of renewable energy their small community of approximately 11,000 might be able to incorporate. One group of six students investigated wind energy and determined through their preliminary poll and informal discussions with city officials that there might indeed be interest in investigating a wind energy project for the town. The students investigated Zond Corporation as the most likely resource for such a project and discovered that in addition to the large-scale wind wall projects like those at Victory Garden and Sky River, Zond also manufactures and manages smaller-scale projects that the students determined might be more appropriate for their community. To begin with, the students wondered if one strategically placed Zond Z-class turbine could provide more efficient and cleaner energy to power the high school and its immediate surrounding facilities (including the football and track stadium).

In their research, students investigated the mechanics of wind power and discovered that before any project could be started, a viable area with "productive winds" must first be located. The students knew that one advantage the Indianola area had was that the terrain was flat and unobstructed. Various farm fields and grazing areas in the county could, in their estimation, allow for "productive" winds. However, the Iowa climate and its distance from significant bodies of water had the potential to counter the positive characteristic of flat terrain. The students determined finally that their findings were limited by the scope of their study. Because of the complexity involved in conducting a complete wind energy assessment study, students concluded their research project with a proposal to bring in an assessment team from Zond Corporation to study the Indianola area in order to determine how productive its winds were.

Examine the tone of Mark Peterson's letter to Jan Mills and her class. Is it encouraging? Does it appear that Zond is interested in the community of Indianola?

In April 1995, Jan Mills's team of science honors students wrote to Zond Corporation's marketing office in Tehachapi, California, asking if a representative or a team of Zond employees would be willing to visit the area and talk with their class about wind power possibilities for Indianola. A Zond public relations officer, Mark Peterson, quickly replied with the letter on pages 36 and 37 addressed to Jan Mills and her students.

Cary Winterbach contacted Jan Mills during the second week in May and the two agreed to a meeting time in October 1995.

"By the way," Mills mentioned at the end of their phone conversation, "Mr. Peterson's letter mentioned that it was up to us whether we wanted the students to participate in a meeting with council members, investors, and landowners."

***Zond Corporation***
Marketing and Development
112 South Curry Street
Tehachapi, CA 93561

Mark Peterson, Public Relations Officer
Office of Public Relations
Bldg. B Office 28
112 South Curry Street
Tehachapi, CA 93561

Ms. Jan Mills and students
Indianola High School
238 North B Street
Indianola, Iowa 50125

April 21, 1998

Dear Ms. Mills:

We received your letter and gracious invitation this week and want to thank you for both. Your students' project sounds very ambitious, and it was gratifying for us at Zond to read of their interest in wind power and Zond in particular. Obviously, your students represent the generation of young adults who will soon be making important decisions regarding national energy uses, and it is encouraging to know that these young people are actively searching for sources of energy with long-term benefits for the planet.

You mentioned in your letter that your school year ends in just five weeks. Because of this, we do not feel we can organize and schedule the kind of comprehensive meeting you are interested in conducting in this period of time. Instead, we'd like to offer an alternative plan to bring a small team to Indianola High School in October or November of your next school year. You mentioned that the students you have participating in this honors course are juniors; therefore, we may still be able to see them in the fall. In addition, we would like to have some time to speak with contacts in the area who may also have an interest in exploring possible applications with Zond. We could keep the meetings separate, if you believe this is preferable, or bring students together with city council people, landowners, and potential investors to give them a sense for what is involved with the entire process. That decision would, of course, be yours.

Ms. Jan Mills and students
page 2

Our plan at this point would be to ask the Midwest Regional Development Officer for Zond, Cary Winterbach of Minneapolis, Minnesota, to contact you and arrange the details for a visit. Our team would likely include Mr. Winterbach in addition to an operations expert and a financial officer, unless you were specifically interested in visiting with someone else from the corporation. While we would view the visit as a preliminary step in determining Indianola's needs and its wind energy potential, we would be happy to tailor our visit to meet the needs of those in the audience, including your students. We would encourage them to develop questions prior to and for the visit.

Since we cannot meet prior to the end of your honors course this term, I do have some suggestions for follow-up research your students may consider doing now or early next fall. First, we have a relatively new World Wide Web site that may be of interest to your students. This site, located at **http://www.Zond.com**, offers some background on Zond and windpower in general. In addition, students may fill out a questionnaire and download facts pertaining to windpower. Accessing our World Wide Web site could provide your students with more of an introduction to Zond and vice versa. In addition, students may also consider looking into your various state resources to determine what is currently available for private energy development. Encourage your students to look into the state energy commission and other related government organizations and determine what processes are necessary for development of this kind. Finally, your students could benefit from investigating the climatological background of Iowa, and specifically the part of the state where Indianola is located. One of the key considerations in locating a windpower generator project is the productivity of the wind for that area. An important piece of information for us would be the average wind speed in the area. These are just a few threads of inquiry your students might consider taking up as we await our meeting in the fall.

Again, thank you for your interest in Zond, and we hope that a fall meeting will be convenient for you. Please anticipate a phone call from Cary Winterbach to negotiate more concrete plans. Please give my best wishes for continued academic success to your students, Ms. Mills.

Best Regards,

Mark Peterson, Public Relations Officer

Note Winterbach's confusion here. What does this indicate?

"I'm not sure what exactly his letter to you entailed, Ms. Mills. I only received a copy of the original letter you sent requesting the meeting. What was the issue?"

"Well, Mr. Peterson suggested that the students could participate in a meeting right alongside others in the community who would be involved in making a decision about whether to begin such a project. I think his letter mentioned inviting city council members and landowners together with the students so that the students could get the big picture on such a project."

How would you define the potential for conflict here?

"I see," replied Winterbach. In fact, town meetings aren't the normal procedure for Zond's projects, even in the planning stages. Zond has traditionally been careful to develop relationships with prospective investors, landowners, and local and state government entities on a one-to-one basis. In fact, numerous investors are reticent to have their identities known in their communities. Knowing this, and Zond's history of individually tailored services to its clients, Winterbach is surprised that Peterson made the offer for this town meeting approach. But he remains quiet about the problems he sees with the idea as he talks with Ms. Mills.

"I thought about Mr. Peterson's suggestion," continued Mills, "and I thought that he sounded like a real teacher. He's right. It *would* be a great learning experience for them. In fact, I mentioned this to the students involved on the project, and they're very excited about the prospect of meeting with others in the community about wind-generated energy. Knowing that city council members and investors might actually be considering this idea lends a lot of credence to the work the students have already done. It makes the whole thing very real for them."

"I agree, Ms. Mills," Winterbach replied. "Sometimes, though, such meetings aren't possible if, for example, prospective investors want to remain anonymous. But I'll look into the possibility of such a meeting."

"Of course I understand that," agreed Mills. "But I'd sure love to give it a try if it's remotely possible. It would be a great learning experience for my students."

The two agree to talk again to finalize details at the end of the summer, and Cary Winterbach resolves to get to the bottom of the town meeting proposal.

## Background Development

How is your role defined here? What is your professional background?

You have been asked to serve as the operations officer (engineer and technical support services representative) for the team of three designated to go to Indianola. In case appendix A, you'll find an e-mail exchange between Cary Winterbach, Midwest regional development officer, and Mark Peterson, public relations officer, which Winterbach has passed on to you. The exchange refines the purpose for what Winterbach calls the "Indianola town meeting." Read through the exchange between the two and analyze what the purpose and content of the informational meeting needs to be. Then, in a brief memo to Winterbach, summarize what you see to be the purpose of the town meeting, provide a list of the main points of information you will plan to cover as the engineering representative for the group, and highlight possible problems you see with any aspect of the meeting or your role in it. You may use any information available to you on the World Wide Web site, in the appendices, or in your own research on Zond.

## Gathering and Reporting Research

As the operations officer representative for the team, you have been called to an informal meeting with Cary Winterbach and Ellen Mayer, a financial officer, to brainstorm and discuss an outline for the meeting in Indianola. Prior to your meeting, though, Cary has asked you to do a little research on the climate and average wind speeds of the Indianola area. Although this research will not be as systematized or reliable as an actual windpower productivity study that Zond would conduct prior to beginning any project, the information will help the team to use actual data as you speak with your participants. Using resources you may find in your library (climatological indexes, almanacs, maps, state statistical data, and energy reports) locate some basic climatological and energy use background on the state and, ideally, Indianola itself. Address the following issues in a brief memo you will hand to Mayer and Winterbach at your meeting.

- What is the state's altitude?
- What is the average wind speed for south central Iowa during the periods of January 1–March 1; March 2–June 1; June 2–September 1; and September 2–December 31?
- What are the average high and low temperatures for the state or for the Indianola area?
- What are the extreme high and low temperatures (records) for the state?
- What is average consumption (in kw/yr) of energy for the state of Iowa? Can the consumption be broken down by energy source? If so, list the energy sources in descending order of most to least used and percentages (if available).
- Does the state currently support any windpower projects (small or large)? If so, is there any information available on where those projects are located?
- Be open to using any other information you uncover in your search that you feel may be helpful background to create a more complete understanding of Indianola or the state of Iowa.

## Summary Task

In addition to presenting an overview of the Z-class turbines themselves, it also falls to you as the engineer to talk about how wind is harnessed to produce energy. While those in your audience may have some background in engineering or energy development, you must tailor your discussion of these topics to the majority in your audience who have little experience or knowledge of such things. Winterbach has encouraged you to focus on summarizing two key concepts for the purposes of this presentation: (1) a summary of how kinetic energy is derived, and (2) a summary of Zond's power output curve information on its generators. In case appendices B and C you will find technical information associated with both. In addition, you may feel free to supplement what is here with your own research. Based on the technical information you uncover, summarize both concepts in a short report. This report will be for your use alone as you develop the oral presentation you will give in Indianola.

## Major Task 1

With your knowledge of the audience you will face, create a 10–12 minute oral presentation that provides an overview of Zond Corporation's windpower generators, the climatological characteristics necessary to run the generators, and the power/output curve provided in case appendix C.

## Correspondence Task

Two days before the presentation in Indianola is scheduled to happen, Cary Winterbach is sent to the hospital with acute appendicitis. Because you are the senior member of the team now, you must decide whether to assume Cary's role in the presentation, which was to provide an introduction to Zond, its mission, and history, or if you should call Tehachapi and ask Mark Peterson to come up and fill in for Cary. If you choose the former, you may have to present both your portion of the meeting (the operations section) and Cary's. While this could look good, it is also a great deal of responsibility. You have been unable to reach Mark Peterson by phone; his secretary has suggested you send him an e-mail with your decision. He will check his e-mail this evening and get back to you first thing in the morning. Compose an e-mail message with your solution to the problem and rationale for the decision you've made.

## Major Task 2

Mark Peterson has told you he will be unable to substitute for Cary Winterbach at the meeting in Indianola, so you have no choice at this point but to present Cary's component of the town meeting. Working from the materials presented here along with the World Wide Web visuals and background information, compose a 10-minute oral presentation that introduces the participants of the meeting to Zond Corporation, its windpower projects, and its mission and history.

# Case Appendix A: Zond Corporation Interoffice E-mail

Examine how the purpose for the town meeting is refined and narrowed here. Discuss this process with your classmates.

Date: Wed, 4 June 1998 13:34:07 -0500
From: cwinterbach (Cary Winterbach)
Subject: Iowa meeting questions
To: mpeterson (Mark Peterson)

Mark,

When I called on May 28, Marta told me you had just left for Seattle. I hope your vacation was relaxing and productive.

I'd like to talk a little bit about this meeting you arranged with Jan Mills in Iowa. When I spoke with her last month, she mentioned that you had suggested a large group setting that would include her students along with local landowners, city council members, and prospective investors to name a few. While it sounds like an interesting idea, I'm not exactly sure what we have in mind here. As you know, this is not how we've traditionally done things. Was this a casual suggestion, or is this a change in practice I have somehow missed?

As I mentioned in our last quarterly meeting, I have been developing relationships with several prospective mid-size investors from the central and south-central Iowa area. While I'm there, I definitely want to meet with these folks; however, I'm not sure what to do about inviting them to a large group meeting that includes the high school students. I need to get a better sense from you about what you have in mind here. Can you give me an outline of your ideas with regard to this town meeting?

Cary

Date: Wed, 4 June 1998 15:15:07 -0500
To: cwinterbach (Cary Winterbach)
Subject: Iowa meeting questions
From: mpeterson (Mark Peterson)

Cary,

I'm sorry to have sprung this on you the way I did. I know I should have called before I left for Seattle, or better yet, just after I'd sent you Jan Mills's letter requesting the meeting. But, as you know, things got away from me.

First off, I know that what you call the "town meeting" approach isn't exactly standard protocol for new development projects. At the same time, we've never had a high school class bring a prospective project to us before, either. When I received the letter from Iowa, I went to Vice President MacDonald and asked what we wanted to do with it. After some discussion, we agreed that this might be an interesting opportunity to experiment with a slightly different format (the larger town meeting approach). Since we don't have a project currently up and running in the area, we thought that the exposure could only do us good.

We anticipated the large group meeting would be an opportunity to educate on a rather general level about Zond, the corporation's mission and history, and the

*(continued on next page)*

*(continued from previous page)*

process involved in determining the viability of an area for a windpower project. It wouldn't be much different from some of the initial meetings you have with investors to talk about electrical generating facilities and Zond's potential in a given area. MacDonald mentioned that it would perhaps be helpful for you to take along an engineer and a finance officer to help present some of the nuts and bolts (so to speak) of the Z-class turbines (e.g., the power output/curve) and costs (e.g., maintenance, start-up, on-going investment and risk). Of course, that would be up to you. It certainly wouldn't come out of your budget if you took a couple of folks with you. Time and structure of the meeting would, of course, also be up to you.

So, does this help? Let me know what you think. Since the duty falls to you to arrange it, this is really your call.

Mark

Date: Thurs, 5 June 1998 10:42:30 -0500
From: cwinterbach (Cary Winterbach)
Subject: RE:Iowa meeting questions
To: mpeterson (Mark Peterson)

Mark,

Thanks for your quick response. While I don't disagree with the general philosophy out of which this idea has come, I'm still unclear about the implementation component of the plan.

I don't mind doing the town meeting if you and MacDonald feel like it might cultivate productive relationships. What I'm still not clear about, however, is how I (or my team) can present information that is at the same time general enough for some eleventh-grade science students, relevant enough for seasoned investors, and technical enough for engineers and/or landowners. These are pretty different audiences, Mark, with very different purposes for being there. Is the idea that the town meeting be a general overview with a participatory follow-up (question and answer)? Then do we plan to have break out sessions with the various groups? I can tell you now that there will be several of the investors I've been talking with who will choose not to be there, if that's the case.

Sorry if I'm sounding dense about this. Just trying to get the whole picture before I set anything up.

Cary

Date: Thurs, 5 June 1998 15:02:17 -0500
To: cwinterbach (Cary Winterbach)
Subject: Iowa meeting questions
From: mpeterson (Mark Peterson)

You don't sound dense. Maybe we should talk about this on the phone so we can iron out any confusion. If, after this e-mail you still want some ideas, give me a call. I'll be here until 6 p.m. my time.

You're right. The audiences are indeed different, with differing needs. But the situation is also unique. Brought together as a group, the idea is to encourage an excitement about the prospects of such a project in Indianola. While I'm not suggesting this be a pep rally, the effects may be similar. Yes, the town meeting would probably work best as a more general overview followed by a question-answer period. I think maybe this will set the stage in terms of more detailed and specialized discussions later in the process, if there is indeed potential for a "later." Think of this as just the first step in looking at the community. I know it's different from talking to investors who are already interested, but I think it'll still feel like a normal early call—just with more people.

As MacDonald and I talked about the possibility of this sort of meeting, we came up with a couple of items we thought would be likely for the team to touch on. You'll ultimately be the better judge of whether these ideas are all important and/or possible in a meeting of this size, but this is what we talked about:

* Zond's history, purpose and mission

* An introduction to the Z-class turbines and the concepts of the windpower projects

* The design of the turbines and the Z-40 power curve/output

* The windpower study

* The time frame involved in beginning a project

* Financial considerations and maintenance issues

You can determine who talks about what if you choose to take others with you, though the lines of responsibility seem pretty clear. Let me know if you want to talk more.

Mark

# Case Appendix B: Definition of Kinetic Energy[1]

## Kinetic Energy

Power available in the wind in a swept area varies with the size of that area, the density of the air, and the square of the velocity. Energy extractable from the oncoming wind stream over a period of time varies as the size of swept area, density and cube of the velocity, as described in the following equation, where the K.E. is the kinetic energy available,

$$\text{K.E.} = kDV^3$$

$k$ is the density of air. $D$ is the sweeping blades' diameter. $V$ is the average wind speed. At sea level, K.E. = 0.000935.

[1] From the McGraw-Hill Encyclopedia of Energy (New York: McGraw-Hill, 1976), p. 734.)

# Case Appendix C: Power Curve Table[1]

Projected energy production is expressed as the number of kilowatt hours per year (kWh/yr) that a turbine is estimated to produce. The estimation of power production involves the application of a turbine's power curve to the wind data from the turbine's proposed location. A turbine's power curve is simply a graph of the turbine's production in kilowatt hours at given wind speeds. Wind data consists of historical data on wind speeds and energy levels over a period of time. A power production estimate combines an average year of wind speeds with the turbine's power curve to arrive at projected output for the year.

The calculation of the estimated power production from the turbines will depend upon their power output at various wind velocities (the "power curve"), the expected wind conditions at the operating site, and the estimated turbine availability rate and power transfer line losses.

Based upon the operating data of Zond's 364 Model V15 turbines and available engineering data, the following table reflects the power curve available for the V15 turbine:

| Wind velocity | | Power output |
|---|---|---|
| 0–8 mph | | 0 kW |
| | Small generator cut-in | |
| 10 mph | | 1.4 kW |
| 15 mph | | 11.8 kW |
| | Switch to large generator | |
| 20 mph | | 29.6 kW |
| 25 mph | | 47.8 kW |
| 30 mph | | 60.9 kW |
| 35 mph | | 68.0 kW |
| 40 mph | | 72.0 kW |
| 45 mph | | 73.0 kW |
| 50 mph | | 74.5 kW |
| 55 mph | | 74.5 kW |
| 60 mph | | 74.5 kW |
| | Generator cut-out | |
| Above 62 mph | | 0 kW |

[1] From the *Zond Victory Garden Management Program Manual*, internal document.

# Summary of Evaluative Criteria for Major Tasks in Case 4

| | ***1 Unacceptable***<br>Insufficient answer to assignment expectations | ***2 Below Average***<br>Inappropriate or ineffective verbal/ visual choices limit document success | ***3 Meets Task Expectations***<br>Has answered objectives of assignment, but individual components could be strengthened | ***4 Above Average***<br>Few flaws, document meets expectations, but could benefit from more attention to detail | ***5 Excellent/ Professional***<br>Few or no flaws, demonstrates keen insight into case subtleties and details |
|---|---|---|---|---|---|
| **Purpose/Key Points**<br>▪ Identifies and meets purpose<br>▪ Articulates key points clearly<br>▪ Demonstrates understanding of technical terms and Zond Corporation | | | | | |
| **Context**<br>▪ Identifies/defines context and situational constraints<br>▪ Demonstrates awareness of colleagues' & audience's needs for situation | | | | | |
| **Audience**<br>▪ Identifies/defines layperson audience's informational needs<br>▪ Establishes appropriate tone<br>▪ Researches and uses technical information effectively<br>▪ Presentation offers appropriate level of technical information | | | | | |
| **Design**<br>▪ Demonstrates awareness of visual design elements of oral presentations<br>▪ Demonstrates an awareness of design options & technological aids in the development of these options | | | | | |

# PART II

# *Developing the Communicator's Tools*

# CASE 5

# *Reporting Test Results: SRI's New Biosyringe*[1]

This case, set in a medical supplies testing lab, asks you to work in two- or three-person teams. Teams are introduced to the biosyringe—a tiny syringe designed to inject individual infected cells. You are asked to analyze a series of charted testing results and report them to the biosyringe test team leader.

## Background

Even by narrowing the focus to living infectious agents, any team would have a great many research methods to choose from. Discuss different research methods with a biological sciences expert to gain insight into the field.

The Seattle Research Institute (SRI) is a large medical testing laboratory located in Seattle, Washington. Recently, SRI and a leading biomedical engineering team reported preliminary lab testing results on something called the Hauch biosyringe. Because the cell is the basic unit of life (and therefore the basic unit of disease) a key medical breakthrough in disease treatment would involve some means of prevention or treatment of individual cellular injury. Cellular injury can be caused by:

- oxygen or other critical nutrient deficiency
- physical disruption
- living infectious agents
- introduction of outside chemical agents (e.g., toxins or drugs)

The Hauch team is most interested in addressing the third bullet—living infectious agents, their effects on cells, and the means by which medical science might directly address and treat specific or individual cells. A breakthrough in terms of treating specific injured cells might have long-term ramifications for treating cancer, tuberculosis, and many other diseases.

[1] The biosyringe in this case is based in part on the Scripps Research Institute's development of the nanosyringe. Specific details of the actual nanosyringe have been altered for the purposes of this case. The author wishes to thank Greg Howe of Lifecore Corp. for his insights into research and data reporting for the "biosyringe" in this case. In addition, many thanks to Melissa Frantz, my friend and sister-in-law, whose nursing and medical research training were also very helpful.

Here you are introduced to the biosyringe.

The Hauch biosyringe, developed by biomedical engineer Kip Hauch, is, in essence, a tiny "pseudo-syringe" built from amino acid rings designed to penetrate individual infected cell walls. The biosyringe was constructed to embed in cell walls in order to inject an individual infected cell with appropriate antibacterial agents. By varying the number and type of amino acids, Hauch and his team were able to control the size and shape of the individual biosyringes and were subsequently able to "program" the biosyringes to seek out the appropriate cells with which to bind.

Clearly, the delivery means is most important here. The treatments for various infectious agents have been around for many years, but treating individual cells is a real advance. Discuss with classmates what other breakthroughs this might prompt.

The team's initial efforts from 1992 to 1995 focused on developing an appropriate simulation test. The testing involved artifical membranes that served as cell walls and simple glucose combinations that served as the faux "drug." After successful testing in the artificial environment, the team focused on developing a means of delivery that would work for animals and humans. Two large teams worked separately on developing a pill that would carry the nanosyringes through the bloodstream via the digestive tract and also through an intraveneous drip method.

You may wish to look up the biomedical terms.

In 1995, researchers took the biosyringe development to the next stage—animal testing. Because researchers were particularly interested in seeing how the biosyringe might work against streptococcus, rats were introduced to a form of *Erysipelas*, a serious streptococcal infection of the skin in which lesions, high fever, and toxic reaction occur. Erysipelas infections were cultivated among rat populations in the lab, and the biosyringes, loaded with antibiotics, were delivered both by pill and through injections. Those rats that responded best were injected with the biosyringe compound (noted as 2g2X). While rats were also given pills, many did not respond as quickly and some died of the infection.

As a result, the teams determined it was important to develop capsules that ensured a faster delivery of the biosyringe. Fast response time was a key criterion in achieving the stage at which the Hauch team might experiment with human subjects, particularly given the interest in working with streptococcus. If, for example, the biosyringe was to be effective in treating the much-publicized flesh-eating streptococcus virus in humans, it must work within hours because of the wildfire rapidity of the bacteria's development in victims' systems.

In 1997, the Hauch research team focused its efforts on developing an appropriate delivery system for the biosyringe to increase positive response time. The team found a small number of rats that did not respond at all to 2g2X (either intraveneously or through the ingested method), and a small part of the team broke off to look into why the rats' systems did not respond. The remaining portion of the team stayed focused on fine-tuning the speed of delivery into the system through the pill.

## The Situation and Your Role

Your role and purpose are defined here.

You are a laboratory specialist on the Hauch team. Along with the rest of the team, you have learned that Hauch and SRI have come under some pressure to publish positive results quickly on this project because some unnamed key funding may dry up. As a result, many team members have been putting in long hours on testing and periodic reporting (even if results feel preliminary for now). The goal is to determine whether the latest capsule has effectively improved delivery time enough to warrant possible human subjects testing—the next big step in the process for moving the biosyringe into actual use.

Your job in this testing stage is to monitor 15 lab rats infected with Erysepilas bacteria. You are monitoring response time to antibiotic treatment using the capsule delivery form of the biosyringe. Each of the specimens you work with has been diagnosed as positive for the infection and has been treated. Your purpose is to determine how effective the biosyringe delivery of antibiotics is. Your job is to monitor lesion development, behavior, and fever; you must also make notes every hour for each rat. Another team is monitoring urine output and white blood cell count. You may assume that your findings will ultimately be combined; right now, however, you are not responsible for the other team's data.

Note the limitations, discussed here, on the data located in the appendix.

Note that in the lesion development section of your appendix, the team uses an internal rating system. A level 1 (or zero in some cases) indicates only minor or no development. A 10 rating would be acute or life threatening. The rates between the two extremes indicate an increasing level of development and seriousness. The body temperature is based on the average rectal temperature for laboratory mice and rats. Normal is approximately 35.5° C under ideal external temperature conditions. The data offered in the behavior section are qualitative and based solely on 5-minute observation increments every hour.

## Background Development

Consider the following questions as a way to fine-tune your understanding of the case and its details. Answer the questions alone or work in small groups of two or three to discuss the answers. Feel free to draw on your responses to inform any of the tasks that follow:

- It would be useful for you to know a little about the streptococcus bacteria and, more specifically, Erysepilas. With a partner or two, go to the library and/or to your biology department chair and gather as much information as you can about the manifestations of the infection. What does it look like? How is it diagnosed? How quickly can it spread in a patient? How is it related to other strep viruses?

- You are asked to report technical information based on the charts provided in the case appendix. Discuss with your partners how best to present this information efficiently and effectively. Take notes on this discussion to use as you develop your presentation of the information.

- How does the pressure of time figure into reporting your findings? Might this time crunch also affect the outcome? Discuss with your class the potential problems this timetable has for you.

- Consider the ramifications if your team does not determine that there is positive proof of the effectiveness of this delivery method. Discuss with your classmates the dangers of this kind of pressure to illustrate appropriate results.

## Major Task 1

Using the notes on the rats you have monitored (located in the case appendix), summarize the progress and response time for each rat, assuming you will use these summaries in a larger report to your supervisor. Make your summaries succinct but complete. Clearly, your main objective is to determine whether you can observe any progress in your rats and at what point that progress is made. Keep in mind, however, that without proof of progress, it is likely your team will lose its funding and your research may be halted.

## Visual Design Task

Using the specific data for each specimen, create a graph that depicts the changes that occur in two of the three variables (for example, lesion development and fever). The graph may be bar or line, and you may feel free to use color or whatever design elements will help you with clarity. You will create 15 graphs (one for each subject) and your purpose will be to compare the two variables visually. This type of visual summary may help you to make conclusions for each specimen as well as for your group as a whole.

## Correspondence Task

Using your summaries from Major Task 1, write an internal memo to Kip Hauch, Project Manager, detailing your findings. Assume that Hauch will require all lab test teams to report similarly. Summarize the test setting, specific data, and conclusions. You may work with a partner on this task, if you wish.

## Major Task 2

Hauch has appreciated your careful attention to detail and accuracy in your reports, and he has asked you to give a brief presentation to the rest of the lab assistants on appropriate reporting techniques. He has complained to you privately that the write-ups he received were often sloppy or indecipherable, both in terms of language and organization. You may use any outside resources that fit your needs, and you may also choose a partner to work with. Create a 15-minute presentation on scientific report writing drawing from conventions in physiology and/or cell biology journals. Emphasize the importance of keeping a publishing objective at the forefront. Also, remember that these people are your colleagues and that they have essentially been told that what they have produced is substandard. Your colleagues are not happy and may be resentful of a presentation style that underscores this message. Be aware of the audience expectations and needs in this situation.

# Case Appendix: Data from Specimen Observation

## Lesion Development

*Specimen 220*

| Observed Hour | Lesion Rating | Notes |
|---|---|---|
| 1 | 0 | No observable lesion development |
| 2 | 2 | Redness and swelling noted in left hindquarter region |
| 3 | 2 | Redness and swelling not observably different |
| 4 | 4 | Surface of skin in affected region broken |
| 5 | 4 | Unchanged |
| 6 | 4 | Unchanged |
| 7 | 4 | Unchanged |
| 8 | 4 | Unchanged |
| 9 | 4.5 | Diameter of lesion increased .20% |
| 10 | 4.5 | Unchanged |
| 11 | 4 | Affected region dried 40% |
| 12 | 3 | Diminished swelling |

*Specimen 221*

| Observed Hour | Lesion Rating | Notes |
|---|---|---|
| 1 | 0 | No observable lesion development |
| 2 | 0 | Unchanged |
| 3 | 0 | Unchanged |
| 4 | 0 | Unchanged |
| 5 | 0 | Unchanged |
| 6 | 0 | Unchanged |
| 7 | 0 | Unchanged |
| 8 | 0 | Unchanged |
| 9 | 0 | Unchanged |
| 10 | 0 | Unchanged |
| 11 | 0 | Unchanged |
| 12 | 0 | Unchanged |

***Specimen 222***

| Observed Hour | Lesion Rating | Notes |
|---|---|---|
| 1 | 2 | Redness and swelling noted in abdominal area |
| 2 | 2 | Unchanged |
| 3 | 2.5 | Surface area increased .10% |
| 4 | 3 | Surface of skin in affected region broken |
| 5 | 3 | Unchanged |
| 6 | 3 | Unchanged |
| 7 | 3 | Unchanged |
| 8 | 3 | Unchanged |
| 9 | 3 | Unchanged |
| 10 | 3 | Unchanged |
| 11 | 3 | Unchanged |
| 12 | 3 | Unchanged |

***Specimen 223***

| Observed Hour | Lesion Rating | Notes |
|---|---|---|
| 1 | 0 | No observable lesion development |
| 2 | 2 | Redness and swelling noted in right hindquarter region |
| 3 | 2 | Unchanged |
| 4 | 2 | Unchanged |
| 5 | 2 | Unchanged |
| 6 | 1 | Diminished swelling |
| 7 | 2 | Redness and swelling noted in left hindquarter region |
| 8 | 3 | Skin surface on left hindquarter broken |
| 9 | 4.5 | Diameter of lesion increased .20% |
| 10 | 4.5 | Unchanged |
| 11 | 4.5 | Unchanged |
| 12 | 5.5 | Diameter of lesion (left hindquarter) increased .15% |

***Specimen 224***

| Observed Hour | Lesion Rating | Notes |
|---|---|---|
| 1 | 0 | No observable lesion development |
| 2 | 0 | Unchanged |
| 3 | 0 | Unchanged |
| 4 | 0 | Unchanged |
| 5 | 0 | Unchanged |
| 6 | 0 | Unchanged |
| 7 | 0 | Unchanged |
| 8 | 0 | Unchanged |
| 9 | 0 | Unchanged |
| 10 | 0 | Unchanged |
| 11 | 0 | Unchanged |
| 12 | 0 | Unchanged |

*Specimen 225*

| Observed Hour | Lesion Rating | Notes |
|---|---|---|
| 1 | 0 | No observable lesion development |
| 2 | 2 | Redness and swelling noted in left hindquarter region |
| 3 | 2 | Redness and swelling not observably different |
| 4 | 4 | Surface of skin in affected region broken |
| 5 | 5 | Diameter of lesion increased .20% |
| 6 | 5 | Unchanged |
| 7 | 6 | Diameter of lesion increased .10% |
| 8 | 6 | Unchanged |
| 9 | 6 | Unchanged |
| 10 | 6 | Unchanged |
| 11 | 7 | Diameter of lesion increased .09% |
| 12 | 8 | Diameter of lesion increased .05% |

*Specimen 226*

| Observed Hour | Lesion Rating | Notes |
|---|---|---|
| 1 | 4 | Lesions noted on abdomen and back areas |
| 2 | 4 | Unchanged |
| 3 | 4 | Unchanged |
| 4 | 3.5 | Surface of skin in affected region dried |
| 5 | 3.5 | Unchanged |
| 6 | 3.5 | Unchanged |
| 7 | 2 | Swelling decreased in both affected areas |
| 8 | 2 | Unchanged |
| 9 | 1.5 | Diameter of lesion in abdominal area decreased .20% |
| 10 | 1.5 | Unchanged |
| 11 | 1 | Diameter of lesion on back decreased .05% |
| 12 | 1 | Unchanged |

*Specimen 227*

| Observed Hour | Lesion Rating | Notes |
|---|---|---|
| 1 | 4 | Lesions on back and neck areas |
| 2 | 5 | Diameter lesion on back increased .05% |
| 3 | 5 | Unchanged |
| 4 | 6 | Diameter lesion on neck increased .10% |
| 5 | 6 | Unchanged |
| 6 | 6 | Unchanged |
| 7 | 5 | Swelling decreased in both affected areas |
| 8 | 5 | Unchanged |
| 9 | 4.5 | Diameter of back lesion decreased .02% |
| 10 | 4.5 | Unchanged |
| 11 | 4.5 | Unchanged |
| 12 | 4.5 | Unchanged |

***Specimen 228***

| Observed Hour | Lesion Rating | Notes |
|---|---|---|
| 1 | 0 | No observable lesion development |
| 2 | 0 | Unchanged |
| 3 | 0 | unchanged |
| 4 | 2 | Redness and swelling noted in left hindquarter region |
| 5 | 2 | Unchanged |
| 6 | 2 | Unchanged |
| 7 | 2 | Unchanged |
| 8 | 2 | Unchanged |
| 9 | 1 | Redness and swelling decreased |
| 10 | 1 | Unchanged |
| 11 | 0 | No observable lesion development |
| 12 | 0 | Unchanged |

***Specimen 229***

| Observed Hour | Lesion Rating | Notes |
|---|---|---|
| 1 | 0 | No observable lesion development |
| 2 | 2 | Redness and swelling noted in back region |
| 3 | 2 | Redness and swelling not observably different |
| 4 | 4 | Surface of skin in affected region broken |
| 5 | 4 | Unchanged |
| 6 | 4 | Unchanged |
| 7 | 4 | Unchanged |
| 8 | 4 | Unchanged |
| 9 | 4.5 | Diameter of lesion increased .20% |
| 10 | 4.5 | Unchanged |
| 11 | 4 | Diminished swelling |
| 12 | 3.5 | Further diminished swelling |

***Specimen 230***

| Observed Hour | Lesion Rating | Notes |
|---|---|---|
| 1 | 0 | No observable lesion development |
| 2 | 0 | Unchanged |
| 3 | 0 | Unchanged |
| 4 | 0 | Unchanged |
| 5 | 0 | Unchanged |
| 6 | 0 | Unchanged |
| 7 | 0 | Unchanged |
| 8 | 0 | Unchanged |
| 9 | 0 | Unchanged |
| 10 | 0 | Unchanged |
| 11 | 0 | Unchanged |
| 12 | 0 | Unchanged |

***Specimen 231***

| Observed Hour | Lesion Rating | Notes |
|---|---|---|
| 1 | 4 | Lesions on abdominal and facial areas |
| 2 | 4 | Unchanged |
| 3 | 5 | Redness and swelling on left hindquarter |
| 4 | 7 | Surface of skin in left hindquarter region broken; diameter of lesion in abdominal area increased .10% |
| 5 | 7.5 | Diameter of lesion in facial area increased .01% |
| 6 | 7.5 | Unchanged |
| 7 | 8 | Lesion in left hindquarter region increased .20% |
| 8 | 8 | Unchanged |
| 9 | 9 | Diameter of lesion in abdominal area increased .20% |
| 10 | 9 | Unchanged |
| 11 | * | Deceased |
| 12 | * | Deceased |

***Specimen 232***

| Observed Hour | Lesion Rating | Notes |
|---|---|---|
| 1 | 0 | No observable lesion development |
| 2 | 2 | Redness and swelling noted in right hindquarter region |
| 3 | 2 | Unchanged |
| 4 | 2 | Unchanged |
| 5 | 1 | Redness and swelling decreased |
| 6 | 1 | Unchanged |
| 7 | 0 | No observable lesion development |
| 8 | 0 | Unchanged |
| 9 | 0 | Unchanged |
| 10 | 0 | Unchanged |
| 11 | 0 | Unchanged |
| 12 | 0 | Unchanged |

***Specimen 233***

| Observed Hour | Lesion Rating | Notes |
|---|---|---|
| 1 | 0 | No observable lesion development |
| 2 | 2 | Redness and swelling noted in right hindquarter region |
| 3 | 2 | Redness and swelling not observably different |
| 4 | 4 | Surface of skin in affected region broken |
| 5 | 4 | Unchanged |
| 6 | 4 | Unchanged |
| 7 | 4 | Unchanged |
| 8 | 4 | Unchanged |
| 9 | 4.5 | Diameter of lesion increased .10% |
| 10 | 4.5 | Unchanged |
| 11 | 4 | Affected region dried |
| 12 | 3 | Diminished swelling |

*Specimen 234*

| Observed Hour | Lesion Rating | Notes |
|---|---|---|
| 1 | 0 | No observable lesion development |
| 2 | 0 | Unchanged |
| 3 | 0 | Unchanged |
| 4 | 0 | Unchanged |
| 5 | 0 | Unchanged |
| 6 | 0 | Unchanged |
| 7 | 0 | Unchanged |
| 8 | 0 | Unchanged |
| 9 | 0 | Unchanged |
| 10 | 0 | Unchanged |
| 11 | 0 | Unchanged |
| 12 | 0 | Unchanged |

*Specimen 235*

| Observed Hour | Lesion Rating | Notes |
|---|---|---|
| 1 | 2 | Redness and swelling noted in back area |
| 2 | 2 | Unchanged |
| 3 | 2 | Unchanged |
| 4 | 4 | Surface of skin in affected region broken |
| 5 | 4 | Unchanged |
| 6 | 4 | Unchanged |
| 7 | 4.5 | Diameter of lesion increased .10% |
| 8 | 4.5 | Unchanged |
| 9 | 4 | Diminished swelling |
| 10 | 4 | Unchanged |
| 11 | 3 | Affected region dried |
| 12 | 2.5 | Diameter of lesion decreased .10% |

# Behavior

*Specimen 220*

| Hour | Notes |
|---|---|
| 1 | Alert, moderate energy |
| 2 | Alert, diminished energy |
| 3 | Unchanged |
| 4 | Lethargic, less responsive |
| 5 | Asleep |
| 6 | Unchanged |
| 7 | Lethargic, unresponsive |
| 8 | Unchanged |
| 9 | Unchanged |
| 10 | Lethargic, moderately responsive |
| 11 | Unchanged |
| 12 | Unchanged |

*Specimen 221*

| Hour | Notes |
|---|---|
| 1 | Alert, moderate energy |
| 2 | Unchanged |
| 3 | Unchanged |
| 4 | Unconscious |
| 5 | Unchanged |
| 6 | Alert, moderate energy |
| 7 | Lethargic, diminished energy |
| 8 | Unchanged |
| 9 | Unchanged |
| 10 | Lethargic, moderate energy |
| 11 | Unchanged |
| 12 | Alseep |

*Specimen 222*

| Hour | Notes |
|---|---|
| 1 | Lethargic |
| 2 | Lethargic, diminished alertness |
| 3 | Unchanged |
| 4 | Lethargic, less responsive |
| 5 | Unconscious |
| 6 | Unchanged |
| 7 | Lethargic, unresponsive |
| 8 | Unchanged |
| 9 | Unchanged |
| 10 | Lethargic, increased alertness |
| 11 | Unchanged |
| 12 | Unchanged |

*Specimen 223*

| Hour | Notes |
|---|---|
| 1 | Alert, moderate energy |
| 2 | Alert, diminished energy |
| 3 | Unchanged |
| 4 | Lethargic, diminished alertness |
| 5 | Unconscious |
| 6 | Unchanged |
| 7 | Lethargic, unresponsive |
| 8 | Unchanged |
| 9 | Unchanged |
| 10 | Lethargic, conscious |
| 11 | Unchanged |
| 12 | Unchanged |

*Specimen 224*

| Hour | Notes |
|---|---|
| 1 | Alert, moderate energy |
| 2 | Alert, diminished energy |
| 3 | Unchanged |
| 4 | Unchanged |
| 5 | Lethargic, conscious |
| 6 | Unchanged |
| 7 | Unconscious |
| 8 | Unchanged |
| 9 | Unchanged |
| 10 | Alert, moderate energy |
| 11 | Unchanged |
| 12 | Alert, improved energy |

*Specimen 225*

| Hour | Notes |
|---|---|
| 1 | Alert, low energy |
| 2 | Conscious, diminished energy |
| 3 | Unchanged |
| 4 | Lethargic, less responsive |
| 5 | Unconscious |
| 6 | Unchanged |
| 7 | Unchanged |
| 8 | Unchanged |
| 9 | Unchanged |
| 10 | Comatose |
| 11 | Unchanged |
| 12 | Unchanged |

*Specimen 226*

| Hour | Notes |
|---|---|
| 1 | Unconscious |
| 2 | Unchanged |
| 3 | Unchanged |
| 4 | Unchanged |
| 5 | Conscious, lethargic |
| 6 | Unchanged |
| 7 | Unconscious |
| 8 | Conscious, lethargic |
| 9 | Unchanged |
| 10 | Alert, moderately responsive |
| 11 | Unchanged |
| 12 | Unchanged |

***Specimen 227***

| Hour | Notes |
|---|---|
| 1 | Unconscious |
| 2 | Unchanged |
| 3 | Conscious, unresponsive |
| 4 | Unconscious |
| 5 | Unchanged |
| 6 | Conscious, lethargic |
| 7 | Unchanged |
| 8 | Unchanged |
| 9 | Unchanged |
| 10 | Unconscious |
| 11 | Conscious, lethargic |
| 12 | Unchanged |

***Specimen 228***

| Hour | Notes |
|---|---|
| 1 | Alert, moderate energy |
| 2 | Alert, diminished energy |
| 3 | Unchanged |
| 4 | Lethargic, less responsive |
| 5 | Unconscious |
| 6 | Alert, diminished energy |
| 7 | Lethargic, less responsive |
| 8 | Unconscious |
| 9 | Alert, improved energy |
| 10 | Alert, improved energy |
| 11 | Alert, moderate energy |
| 12 | Unchanged |

***Specimen 229***

| Hour | Notes |
|---|---|
| 1 | Alert, moderate energy |
| 2 | Alert, diminished energy |
| 3 | Unchanged |
| 4 | Lethargic, less responsive |
| 5 | Unconscious |
| 6 | Unchanged |
| 7 | Lethargic, unresponsive |
| 8 | Unchanged |
| 9 | Unchanged |
| 10 | Lethargic, moderately responsive |
| 11 | Unchanged |
| 12 | Unchanged |

*Specimen 230*

| Hour | Notes |
|---|---|
| 1 | Alert, moderate energy |
| 2 | Alert, diminished energy |
| 3 | Unchanged |
| 4 | Unconscious |
| 5 | Alert, moderate energy |
| 6 | Alert, diminished energy |
| 7 | Unconscious |
| 8 | Unchanged |
| 9 | Alert, moderate energy |
| 10 | Unchanged |
| 11 | Unchanged |
| 12 | Unchanged |

*Specimen 231*

| Hour | Notes |
|---|---|
| 1 | Unconscious |
| 2 | Unchanged |
| 3 | Unchanged |
| 4 | Unchanged |
| 5 | Unchanged |
| 6 | Unchanged |
| 7 | Comatose |
| 8 | Unchanged |
| 9 | Unchanged |
| 10 | Unchanged |
| 11 | * |
| 12 | * |

*Specimen 232*

| Hour | Notes |
|---|---|
| 1 | Alert, moderate energy |
| 2 | Alert, diminished energy |
| 3 | Unchanged |
| 4 | Lethargic, less responsive |
| 5 | Unconscious |
| 6 | Alert, moderate energy |
| 7 | Lethargic, improved energy |
| 8 | Unchanged |
| 9 | Unchanged |
| 10 | Unconscious |
| 11 | Alert, normal |
| 12 | Unchanged |

*Specimen 233*

| Hour | Notes |
|---|---|
| 1 | Alert, moderate energy |
| 2 | Alert, diminished energy |
| 3 | Unchanged |
| 4 | Lethargic, less responsive |
| 5 | Unconscious |
| 6 | Unchanged |
| 7 | Lethargic, unresponsive |
| 8 | Unchanged |
| 9 | Unchanged |
| 10 | Lethargic, moderately responsive |
| 11 | Unchanged |
| 12 | Unchanged |

*Specimen 234*

| Hour | Notes |
|---|---|
| 1 | Alert, moderate energy |
| 2 | Alert, diminished energy |
| 3 | Unchanged |
| 4 | Unconscious |
| 5 | Unchanged |
| 6 | Alert, improved energy |
| 7 | Alert, diminished energy |
| 8 | Unchanged |
| 9 | Unconscious |
| 10 | Lethargic, moderately responsive |
| 11 | Alert, moderate energy |
| 12 | Alert, improved energy |

*Specimen 235*

| Hour | Notes |
|---|---|
| 1 | Alert, moderate energy |
| 2 | Alert, diminished energy |
| 3 | Unchanged |
| 4 | Lethargic, less responsive |
| 5 | Unconscious |
| 6 | Alert, less responsive |
| 7 | Lethargic, unresponsive |
| 8 | Unchanged |
| 9 | Unconscious |
| 10 | Lethargic, moderately responsive |
| 11 | Unchanged |
| 12 | Alert, diminished energy |

# Body Temperature

***Specimen 220***

| Hour | Temperature, C° |
|---|---|
| 1 | 36.7° |
| 2 | 37.1° |
| 3 | 37.8° |
| 4 | 38.0° |
| 5 | 38.3° |
| 6 | 38.3° |
| 7 | 38.1° |
| 8 | 38.1° |
| 9 | 37.5° |
| 10 | 37.5° |
| 11 | 37.5° |
| 12 | 37.0° |

***Specimen 221***

| Hour | Temperature, C° |
|---|---|
| 1 | 36.7° |
| 2 | 37.1° |
| 3 | 37.8° |
| 4 | 38.0° |
| 5 | 38.0° |
| 6 | 38.0° |
| 7 | 38.5° |
| 8 | 38.0° |
| 9 | 37.5° |
| 10 | 37.5° |
| 11 | 37.2° |
| 12 | 37.0° |

***Specimen 222***

| Hour | Temperature, C° |
|---|---|
| 1 | 36.7° |
| 2 | 37.1° |
| 3 | 37.8° |
| 4 | 38.0° |
| 5 | 38.3° |
| 6 | 38.3° |
| 7 | 38.1° |
| 8 | 38.1° |
| 9 | 37.5° |
| 10 | 37.5° |
| 11 | 37.5° |
| 12 | 37.0° |

*Specimen 223*

| Hour | Temperature, C° |
|---|---|
| 1 | 38.0° |
| 2 | 38.3° |
| 3 | 39.0° |
| 4 | 39.0° |
| 5 | 39.3° |
| 6 | 39.6° |
| 7 | 38.4° |
| 8 | 37.0° |
| 9 | 37.0° |
| 10 | 36.2° |
| 11 | 36.5° |
| 12 | 36.2° |

*Specimen 224*

| Hour | Temperature, C° |
|---|---|
| 1 | 38.0° |
| 2 | 38.4° |
| 3 | 39.3° |
| 4 | 39.7° |
| 5 | 40.1° |
| 6 | 39.0° |
| 7 | * unavailable |
| 8 | 38.7° |
| 9 | 37.5° |
| 10 | 37.5° |
| 11 | 37.5° |
| 12 | 37.0° |

*Specimen 225*

| Hour | Temperature, C° |
|---|---|
| 1 | 37.7° |
| 2 | 37.9° |
| 3 | 39.2° |
| 4 | 40.0° |
| 5 | 40.3° |
| 6 | 41.0° |
| 7 | 41.0° |
| 8 | 39.9° |
| 9 | 38.2° |
| 10 | 38.2° |
| 11 | 40.1° |
| 12 | 39.9° |

***Specimen 226***

| Hour | Temperature, C° |
|---|---|
| 1 | 39.7° |
| 2 | 39.1° |
| 3 | 39.8° |
| 4 | 39.0° |
| 5 | 38.3° |
| 6 | 38.0° |
| 7 | 37.1° |
| 8 | 37.1° |
| 9 | 36.5° |
| 10 | 36.5° |
| 11 | 36.5° |
| 12 | 36.0° |

***Specimen 227***

| Hour | Temperature, C° |
|---|---|
| 1 | 38.7° |
| 2 | 39.1° |
| 3 | 40.8° |
| 4 | 41.0° |
| 5 | 40.3° |
| 6 | 38.3° |
| 7 | 38.1° |
| 8 | 38.1° |
| 9 | 37.5° |
| 10 | 37.5° |
| 11 | 37.5° |
| 12 | 37.0° |

***Specimen 228***

| Hour | Temperature, C° |
|---|---|
| 1 | 39.7° |
| 2 | 38.1° |
| 3 | 37.8° |
| 4 | 38.0° |
| 5 | *unavailable |
| 6 | 37.3° |
| 7 | 38.1° |
| 8 | 38.1° |
| 9 | 37.5° |
| 10 | 36.5° |
| 11 | 36.5° |
| 12 | 36.0° |

*Specimen 229*

| Hour | Temperature, C° |
|---|---|
| 1 | 37.7° |
| 2 | 38.1° |
| 3 | 38.8° |
| 4 | 39.2° |
| 5 | 39.3° |
| 6 | 39.3° |
| 7 | 39.5° |
| 8 | 40.1° |
| 9 | 40.5° |
| 10 | 40.5° |
| 11 | 39.5° |
| 12 | 39.0° |

*Specimen 230*

| Hour | Temperature, C° |
|---|---|
| 1 | 36.7° |
| 2 | 37.1° |
| 3 | 37.8° |
| 4 | 38.0° |
| 5 | 38.3° |
| 6 | 36.5° |
| 7 | 36.2° |
| 8 | 36.0° |
| 9 | 35.5° |
| 10 | 35.5° |
| 11 | 35.3° |
| 12 | 35.2° |

*Specimen 231*

| Hour | Temperature, C° |
|---|---|
| 1 | 39.7° |
| 2 | 40.1° |
| 3 | 40.8° |
| 4 | 41.0° |
| 5 | 42.3° |
| 6 | 42.3° |
| 7 | 42.1° |
| 8 | 42.1° |
| 9 | 42.5° |
| 10 | 42.5° |
| 11 | * |
| 12 | * |

*Specimen 232*

| Hour | Temperature, C° |
|---|---|
| 1 | 36.7° |
| 2 | 37.1° |
| 3 | 37.8° |
| 4 | 38.0° |
| 5 | 38.9° |
| 6 | 38.3° |
| 7 | 38.1° |
| 8 | 38.3° |
| 9 | *unavailable |
| 10 | 37.4° |
| 11 | 37.5° |
| 12 | 37.0° |

*Specimen 233*

| Hour | Temperature, C° |
|---|---|
| 1 | 38.7° |
| 2 | 39.1° |
| 3 | 38.8° |
| 4 | 37.0° |
| 5 | 37.3° |
| 6 | 37.3° |
| 7 | 37.1° |
| 8 | 37.1° |
| 9 | 36.5° |
| 10 | 36.5° |
| 11 | 36.5° |
| 12 | 36.0° |

*Specimen 234*

| Hour | Temperature, C° |
|---|---|
| 1 | 36.7° |
| 2 | 37.1° |
| 3 | 37.8° |
| 4 | 36.2° |
| 5 | 36.0° |
| 6 | 35.3° |
| 7 | 35.5° |
| 8 | * unavailable |
| 9 | 35.5° |
| 10 | 35.7° |
| 11 | 35.5° |
| 12 | 35.0° |

***Specimen 235***

| Hour | Temperature, C° |
|---|---|
| 1 | 39.7° |
| 2 | 38.1° |
| 3 | 38.8° |
| 4 | 38.0° |
| 5 | 37.3° |
| 6 | 37.3° |
| 7 | 38.0° |
| 8 | 38.1° |
| 9 | 37.5° |
| 10 | 37.5° |
| 11 | 36.5° |
| 12 | 36.0° |

# Summary of Evaluative Criteria for Major Tasks in Case 5

| | ***1 Unacceptable***<br>Insufficient answer to assignment expectations | ***2 Below Average***<br>Inappropriate or ineffective verbal/ visual choices limit document success | ***3 Meets Task Expectations***<br>Has answered objectives of assignment, but individual components could be strengthened | ***4 Above Average***<br>Few flaws, document meets expectations, but could benefit from more attention to detail | ***5 Excellent/ Professional***<br>Few or no flaws, demonstrates keen insight into case subtleties and details |
|---|---|---|---|---|---|
| **Purpose/Key Points**<br>▪ Identifies and meets purpose<br>▪ Articulates key points clearly<br>▪ Demonstrates careful analysis of data<br>▪ Demonstrates appropriate use of technical language skills and design elements | | | | | |
| **Context**<br>▪ Identifies/defines situational constraints<br>▪ Demonstrates communication skills in collaborative partnership | | | | | |
| **Audience**<br>▪ Identifies/defines audience and meets identifiable needs<br>▪ Establishes appropriate tone<br>▪ Is sensitive to colleagues' attitudes in presentation | | | | | |
| **Design**<br>▪ Demonstrates awareness of visual design elements of task<br>▪ Demonstrates an awareness of design options & technological aids in the development of these options | | | | | |
| **Oral Presentation**<br>▪ Is clear and offers appropriate insight into task<br>▪ Demonstrates knowledge of audience and purpose | | | | | |

CASE 6

# Locating and Recording Background Material on an HIV/AIDS Research Project

ChemLabs is a new research and pharmaceutical facility located in Greensboro, North Carolina focused on women's health products. Two new chemists have an interest in AIDS research—particularly as it pertains to women's health—but have been largely secretive about what they have accomplished in their research to date. In research assistant teams of two to three, you are asked to research and report on the U.S. patent and trademark process as well as on the histories and the effectiveness of the nine drugs currently approved by the FDA to serve as background material for the chemists' report.

## Background

The first two paragraphs provide insight into the history of ChemLabs/Prouse Institute. Examine these for insight into the relationship of pharmaceutical companies to research institutions. What can you learn here? What questions are still left unanswered?

Look up RTI in a medical journal or medical dictionary and write down its technical meaning. How can you rewrite the definition in layreader terms?

The ChemLabs/Prouse Institute, built in 1994 mostly from private donations, is a medical research and pharmaceutical facility located in Greensboro, North Carolina. Expanding their philanthropic endeavors to increase breast cancer awareness, millionaires Hugo and Georgianna Van Maanen initiated a development campaign in 1989 for funds to build the biggest medical research institute in the nation with a focus on women's health. Their interest in women's health issues was spurred largely by the 1976 death of their daughter, Margaret Prouse, from breast cancer. The campaign was successful, and in a little more than three years all funds, including endowment monies, were in place or guaranteed.

Early in the development process, the board of directors (which included the Van Maanens as cochairs) entertained a proposal from the existing ChemLabs in Greensboro. ChemLabs was then a private pharmaceutical company unaffiliated with a corporation or conglomerate that produced a wide array of cancer and hormone management drugs. In addition, ChemLabs held the distinction of being the only privately owned pharmaceutical company in the nation to own the licensing rights to a relatively new Reverse Transcriptase Inhibitor (RTI), n4d, of the nevirapine family. This RTI is used as part of the combination therapy for the control of HIV/AIDS symptoms. The drug brought in well over $114 million for ChemLabs in the previous year.

At the time the ChemLabs proposal was brought before the Prouse Institute board, ChemLabs management had a strong desire to expand and was considering affiliating with a large international corporation. However, licensing agreements would be affected by such an affiliation, and owners strongly desired to remain independent. An affiliation with the Prouse Institute, however, was an attractive alternative, and one that to that point in the process no one had considered. If the Prouse Institute board would take on ChemLabs as the pharmaceutical component to the larger project, ChemLabs could feasibly expand independently. In addition, the existing ChemLabs facility was not only excellent for its size but strategically located in an area ripe for development. The ChemLabs/Prouse Institute union was logical for all concerned and ultimately would serve to save on the Prouse Institute projections for development needs.

Because of the union with ChemLabs, the Prouse Institute board needed to modify the research institute's emphasis. Since ChemLabs was already invested in pharmaceutical products that were not necessarily directly or exclusively related to women's health issues, Prouse board members felt that research projects at the Institute could *privilege*, but not be exclusive to, women's health concerns. Such limitations, several donors and board members argued, could preclude positive growth in any number of infectious disease research projects and would limit already successful products at ChemLabs. Clearly, their concern was for ChemLabs' license for the lucrative RTI. The Prouse board agreed to loosely define "health issues" to allow for research in areas that affected both men and women. The Van Maanens agreed to open research project proposals but stipulated a preference for those that focused on issues related exclusively to women's health.

## The Situation and Your Role

Since late 1994, the ChemLabs/Prouse Institute has been developing an array of research teams and products. Among the most active and productive teams have been those that have focused on Huntington's disease, osteoporosis, skin cancer, and breast cancer. One team of breast cancer researchers made news in 1997 with the discovery of genetic chains that possibly indicate an "immunity" to the disease. ChemLabs has continued to successfully market its HIV/AIDS nevirapine drug.

In 1995, the Prouse Institute hired two young chemists to head the HIV/AIDS research project—Li Yuan, a chemist and recent graduate of Harvard Medical School, and Graham Fraanken, an immunologist and researcher at the University of Amsterdam. Yuan and Fraanken had previously collaborated on a number of projects involving children and AIDS. Their ambitious proposal for the Prouse Institute focused on investigating HIV pathogenesis and the immune system to develop combination antiviral therapies designed to ultimately eradicate HIV from an infected person while simultaneously initiating immunoreconstitution. Following the lead of Louise Market, an immunologist from Duke University Medical Center, and pediatrician Richard Hong of the University of

Vermont, Yuan and Fraanken have focused on drug treatment of the thymus, a ductless gland known to play some part in disease resistance in children.[1]

Yuan and Fraanken have been largely secretive about what they have accomplished in their research to date. While there is a good deal of speculation that Yuan and Fraanken have recently hit upon something significant, they and their team have remained "mum" about what it might be. This is not particularly unusual in the medical field because research and research funds are extremely competetive. Funding and patenting frequently depend on *timing*, so if researchers are making gains or discoveries, it is imperative that they keep these gains to themselves until they can release the information publicly. Thus, testing for a given project must be relatively complete. If information about a research team's progress were, for example, to leak out prior to its ability to release results, and another team—perhaps one with more funding or better facitilites—were to latch on to those preliminary results, it is possible that the second team may be able to speed the process and reap the rewards of the "discovery." Unethical as it may be, such scenarios have happened before.

You belong to a team of laboratory assistants at the ChemLabs/Prouse Institute and have been running laboratory tests on mice for the neurology team on the Huntington's disease project. Your job has been to chart changes in behavior across three separate control groups of animals and to provide weekly reports to the research project leaders. Though the progress with the lab testing has largely been satisfactory, positive changes have also slowed considerably in the last six weeks. Your project leaders have conferred and determined that if more significant change is not noted in the next month to six weeks, they will refocus their efforts on another component of an earlier finding.

Your role is defined here. What are its constraints and freedoms in directing the process you are about to undertake?

You are not surprised, then, when your supervisor, Lynn Richards, comes to you with a request, "Since things have slowed down a little around here, Yuan and Fraanken were wondering if we could spare you for a little fast research for them," she tells you. "Are you up for taking some time off over here and helping them out?"

You shrug your shoulders noncommittally and reply, "I could probably spare the time. What are they after?"

Lynn shakes her head. "I'm not exactly sure, but I don't think it involves lab time. I think they're after some fast library research. All I know is that they seem to be getting ready to write something up for publication, and they need some background research for the report. Apparently, they can't spare any of their own research assistants over there—they seem to be working around the clock. They've asked if you'll help them out, and they're also hitting up the Carter-Vega-Richman team for two assistants."

"Well, there isn't much going on around here," you sigh. "But I don't want to leave these guys unattended," you gesture to the cages.

Richards nods. "I know. I'll take care of both yours and mine while you're helping out Yuan and Fraanken."

[1] Background information on current HIV/AIDS research from Elizabeth Pennisi and Jon Cohen, "Eradicating HIV from a Patient: Not Just a Dream?" *Science* 272:8 (June 1996): 1994.

You can see that Richards seems somewhat anxious for you to help the HIV/AIDS team out. They must be on the verge of something relatively significant, which could certainly be important to the institute as a whole. You agree to meet with Yuan later that afternoon.

You are asked to work collaboratively with people you do not know. What sorts of information will be useful for you to gather about your partners? Make a list of questions to ask them.

Note that the purpose of your research is defined here.

At your meeting with Yuan, you are introduced to two other research assistants who will be helping with your project. Yuan tells you that the team is on the verge of publishing some significant findings. In addition, the team, working with ChemLabs chemists, is developing a drug they hope to patent. Thus, Yuan tells you, the team has two specific needs prior to publication. "First, we need one or two of you to do a little background research on the available reverse transcriptase inhibitors (like AZT, BioChem Pharma's 3TC, and Bristol-Meyers Squibb's d4T). Don't worry about giving us information on our own n4d, because obviously we already have that stuff."

"By 'background information,'" you interrupt, "what exactly are you after?"

What skills is Yuan relying on you for?

"I was getting to that," she nods. "First off, we simply need to have the names of the RTIs that are currently available and on the market, their manufacturers, and—if you can find them—the most recent market sales figures. I want a brief description of the drug's history and effectiveness—no more than a paragraph each. This information is strictly background. Second, we need the same information on the available protease enzyme inhibitors. Start with Hoffman-LaRoche's saquinavir since that's the biggest name. Again, in this case we simply want the basics for information—manufacturers, market sales, and a brief history of its effectiveness."

Yuan pauses to allow you to write some notes. "You can put this information into a table or however it is most expedient to communicate the information. Then I need a brief technical description of what a protease inhibitor and a reverse transcriptase inhibitor are."

Here your purpose is narrowed. How?

All three of you look quizzically at Yuan because her last request is somewhat unusual. She explains, "We are going for two separate but related publications and we want them out at approximately the same time. The first is to the *American Journal of Medicine.* Obviously, in that case, we don't have to get into fundamentals. But the second, *Scientific American*, aims at an educated layperson readership. We need the RTI and PI descriptions for that one.

"The other thing we need is a brief report on the U.S. patent and trademark process. This will be specifically a report to myself and Fraanken, of course—not for public consumption. We just want a brief overview of the process, based on what you can find in the library. We haven't gone through it ourselves yet, and although folks over in ChemLabs can help, we want our own reference sheet. Any questions?" she concludes.

"How soon do you need the information?"

Note the limitations imposed by time constraints.

She smiles, "Oh, yesterday. We'll take it as soon as we can get it, but ideally we want it inside a week. I don't anticipate it'll take you too long, despite the fact that you haven't been working with the HIV/AIDS stuff most recently. If you feel like you need more help, send me an e-mail and I'll rustle up a fourth person for you. Otherwise, divide the work and get it to me as soon as you can."

Yuan's pager goes off, and she is forced to leave the meeting.

## Background Development

Consider the following questions as a way to fine-tune your understanding of the case and its details. Answer the questions alone or work in small groups of two or three to discuss the answers. Feel free to draw on your responses to inform any of the tasks that follow.

- This is the second case in this book where you find yourself working within a collaborative group. You also have the choice to expand the group or keep it to three members (see the Correspondence Task later). Think about the effectiveness of larger versus smaller groups in your previous experiences. Which would you prefer and why? Discuss with your team members.
- This case is based on reporting technical scientific information and uses some scientific jargon. How important is it for you in your capacity in this case to understand all of the language or the actual scientific discoveries? Discuss with your classmates.
- Yuan never tells you or your collaborators what it is she and Fraanken and their team is about to report. Is that a necessary piece of information, given your role? Does it make a difference to you and your fellow research assistants that she chooses not to tell you? Why do you suppose she makes this choice?
- Because this case relies heavily on research capabilities, it might be useful to do a sweep of the library resources to discover what is available on the technical information that Yuan requests of you. Divide responsibilities among three- or four-member teams. Ask one person to make an inventory of medical and or scientific journals in your library. Ask another to review books and/or articles that focus on AIDS research. A third person can do an Internet search that focuses specifically on cataloging websites devoted to AIDS research. A fourth might run a database search for available resources (e.g., useful videotapes) through interlibrary loan. Come together after your respective searches and pool your knowledge. Talk about how to narrow the scope or focus on the sources you've uncovered.
- After a brief overview of scientific journals, what citation style do they employ?

## Correspondence Task 1

Discuss with your assigned partners Yuan's offer to add a fourth person to your research team. Do you feel you need one, or can you more efficiently complete the task she has given you with only three people? After this discussion, compose a memo (ostensibly an e-mail) to Yuan and Fraanken explaining your response to the offer. Also confirm in the memo how the research tasks have been delegated among you. This should be a brief memo and serve only as a postscript to your meeting with Yuan. If you feel as though you have unanswered questions about the task, you may also choose to address those in the memo.

## Major Task 1

As a team, delegate responsibilities for research and write up your findings informally as an internal group report. Discuss any gaps you discover in the research you've done, and plan for ways to answer any questions that are perhaps not yet answered in your initial research. Plan how you will present all of the information as a package to Yuan and Fraanken and organize your findings.

## Major Task 2

Finalize your report collaboratively, and create a short cover memo for Yuan and Fraanken explaining its organization and the scope of the information you've provided.

## Follow-up Task

Yuan and Fraanken have reviewed your report and like the work you have done. They have asked you to revise the section on the RTI and PI backgrounds so that the information is limited to one page for quick reference. Discuss the possibilities for revision with your team members and revise.

# Summary of Evaluative Criteria for Major Tasks in Case 6

| | ***1 Unacceptable***<br>Insufficient answer to assignment expectations | ***2 Below Average***<br>Inappropriate or ineffective verbal/ visual choices limit document success | ***3 Meets Task Expectations***<br>Has answered objectives of assignment, but individual components could be strengthened | ***4 Above Average***<br>Few flaws, document meets expectations, but could benefit from more attention to detail | ***5 Excellent/ Professional***<br>Few or no flaws, demonstrates keen insight into case subtleties and details |
|---|---|---|---|---|---|
| **Purpose/Key Points**<br>▪ Identifies and meets purpose<br>▪ Articulates key points clearly<br>▪ Demonstrates appropriate use of technical language skills and design elements | | | | | |
| **Context**<br>▪ Identifies/defines situational constraints<br>▪ Demonstrates understanding of trademark process<br>▪ Demonstrates understanding research strategies | | | | | |
| **Audience**<br>▪ Identifies/defines audience and meets identifiable needs<br>▪ Establishes appropriate tone | | | | | |
| **Design**<br>▪ Demonstrates awareness of visual design elements of task<br>▪ Demonstrates an awareness of design options & technological aids in the development of these options | | | | | |

**CASE 7**

# *Determining Need: Revising Truman County's General Assistance Application*

This case asks you to review an existing document used for processing applications for county financial assistance. Understanding the context for the document is extremely important as you determine what needs to be revised. You face length and time constraints, technological challenges, and interpersonal communication issues as you reorganize technical details both visually and verbally.

## Background

Consider the numbers of people who apply for government assistance in Truman County. Do the math and determine how many Truman County residents regularly fill out assistance applications.

Truman County, Mississippi, has a population of just over 35,000 residents. Its county seat, Summer Valley, claims one-third of the county population at 12,167 residents. Just under 5 percent of Truman County's population regularly applies for some sort of government living assistance (ADC benefits, food stamps, Title XIX, Veterans' Assistance, and Medicaid).

In the last election Truman County voters replaced four of the seven people holding seats in the County Board of Supervisors, in part because of voters' anger at some mismanagement and misappropriation of county funds. The new Board of Supervisors members were elected largely because of their campaigns, which promised more attention to social issues and residents' concerns.

The purpose of the Government Assistance Fund is defined here.

Shortly after the election, Jan Thomas and George Jackson, two of the new Board of Supervisors members, called for a county support network to help short-term financial needs for county residents. In addition to several assistance and educational programs, the plan included the development of a General Assistance Fund (GAF). With the help of several federal grants and local development efforts, Truman County amassed an impressive $300,000 to supplement families in need of assistance. The county plans to continue its efforts to raise money for GAF.

Eligibility is important to understand in dealing with assistance programs. Review the qualifications for GAF here.

Families of two or more people may apply for GAF funds to supplement living expenses; however, there are several stipulations for eligibility. First, families must agree to a repayment of the GAF by actual reimbursement or by at least one family member's participation in the Community Work Experience Program, a community service/community education program designed to supplement job skills training. Second, the county will only fund applicants for up to three months out of every year and only if they are ineligible for state or federal assistance. This usually means that applicants are in need of assistance but that they are employed and earn too much money for state or federal aid. Finally, applicants must provide evidence that the funds provided by the county are spent on living necessities including rent, utilities, food, medications/health care, transportation expenses, or clothing.

Problems with some aspects of the GAF are outlined here.

GAF funds have been in place now for two years, and the program is regularly referred to by county officials as the "poor fund." Approximately 90 applications for GAF assistance have been approved since the program's inception, and county officials have quietly acknowledged that the reimbursement component of GAF is a failure. Only 10 percent of the funds allocated by the county have been repaid in any way. This is not a surprise to county officials, however, and GAF funds are technically written off as loss. While officials still maintain the importance of repayment to the fund, they also acknowledge that most of the applicants are simply in no position to repay the money. Though the repayment plan for GAF funds has largely been unsuccessful, county officials have seen a better success with the work agreement efforts. In the past year, county officials have seen 35 percent of GAF recipients make efforts to serve in the Community Work Experience Program.

## The Situation and Your Role

Your role is defined here. What do you anticipate doing in this role?

You work in the Truman County courthouse in the Family Assistance Office. As GAF assistant director, your job is to process all GAF applications as well as interview applicants and help with fundraising efforts and grant writing to maintain annual resources.

One day a tired-looking woman enters your office and hands you a completed GAF application. "I'd like to set up an appointment to talk with you about this, " she says simply.

You introduce yourself and offer to talk with her now. You ask for a moment to scan the application before you proceed, however, and she nods.

The woman's name is Belinda Johnson. As far as you can tell from the application, the woman has four children ranging in age from 2 to 12 years, and no source of income.

Note the problem that Belinda Johnson has with the GAF form.

When you look up from the application, the woman says, "You know, there's a whole lot you can't understand from that skimpy application form. You can't know that my husband died of a massive heart attack two months ago. He worked hard and brought in enough money for us to live okay, and we decided after our second son, Jules, that I would stay home and tend the children. When he died, I found out he'd cashed in his life insurance policy five or six years ago. So we had nothing." You can hear the anger and sadness in the woman's voice and you listen and nod sympathetically. She continued, "That application also doesn't tell you that I've been looking for work ever since right after the funeral, and I've had a couple of good interviews, but nothing has panned out yet. I don't want to be on assistance. I've never taken a handout in my life. But I don't know what else to do."

You assure her that she seems to be a fine candidate for GAF assistance, and that many people apply who are in immediate need, just like her. This seems to help ease some of her tension. Because you are sensitive to her plight, you reassure her that there are lots of options to help her with her job search, too. You call Jo at Job Service and ask if someone could talk with Mrs. Johnson about job retraining programs and educational assistance. Jo agrees to a 10:30 meeting, and Mrs. Johnson brightens at the prospect.

"You'll be hearing from us by next Monday or Tuesday," you tell Mrs. Johnson, as she stands to leave.

"Thank you for your help," she says. "You know, one more thing about that application form. It made me feel like a criminal as I was filling it out."

You nod and carefully reply, "It may not be the best application form, but could part of that have been because of how you said you feel about accepting a 'handout'?"

She nods, "Yes, I suppose so. And I probably need to get over that feeling so that I can get my family out of this financial mess." With that, she leaves your office.

In your review of the document you note problems with the form as well.

You have heard this complaint about the application from many clients before, but something about Mrs. Johnson's interview sticks with you for the rest of the day. Late that afternoon, you look at her application again and you notice really for the first time how impersonal the form actually is. As Mrs. Johnson pointed out, there is no space for the applicant to explain extenuating circumstances like the one she is facing with the death of her husband. In addition, the tone of some of the sections, particularly the certification statement, could be read by the applicant as somewhat callous or unfeeling.

Your purpose is defined here. What potential obstacles do you face by attempting a revision of the form?

You decide that perhaps the application form could stand to be revised, but you will need approval from the GAF director before you make any changes. In addition, in your search for the original document on the hard drive of your computer, you discover that there is no copy to be found. You surmise this means that the original was likely done on an old word processor or perhaps even an advanced electronic typewriter. The GAF director, Tonia Farney, has headed the program since its inception two years ago and in fact designed the application herself. While you and Tonia generally get along quite well, you have been with the office only six months and you are the third assistant director who has worked in the office. From what you have gathered in conversations with Tonia and with your secretary, Sarah Jessup, the first two assistants didn't work out because of "personality conflicts" with Tonia and because they had a tendency to "overstep their bounds." In your short amount of time in the office, you have found Tonia to be generally open to new ideas, unless they are overtly critical of steps she has taken previously. She is also somewhat autocratic in her methods.

## Background Development

Consider the following questions as a way to evaluate what you've read in the case. Answer the questions on paper or in an online discussion, or work in a group and discuss possible answers orally. You may use your responses to help you in subsequent tasks.

- Can job applications, financial assistance forms, bank loan applications, or citizenship reports communicate *attitudes* about the audience? Find an example of an application or form that seems to communicate an attitude about its prospective user. In a paragraph, briefly describe the form's tone and what evidence you find to support your claim.
- How can visual design choices help the author's attempt to establish tone? Look at the design choices in the form in the case appendix. What do all caps communicate? Boldface? Make a list of design choices you believe work in this form. In a second column compile a list of those design choices you believe are ineffective.
- In a group of three or four, discuss the risks in revising work originally done by a superior. What are some persuasive methods you could use to encourage the superior to see the need for change?

## Getting Started

Write a memo to Tonia proposing to revise the GAF application form. Keep in mind that Tonia is the person who created the first GAF form, and it is only about two years old. You may feel free to suggest alternative access and storage of the form as well. For example, given that the original copy of the document is only available in hard copy, this might be a good opportunity to suggest not only a new version of the application, but also perhaps electronic *access* and record keeping (e.g., online storage of applications). Consider a variety of options that might enhance office productivity as well as reaching the intended audience more effectively.

## Major Task

Tonia has approved your proposal somewhat grudgingly in a weekly staff meeting with you. She notes that the revised form cannot be any longer than the current one (meaning that it should span one page front and back) and that none of the requirements for information can be negated. She is somewhat skeptical that you can revise the GAF form much, given the Board of Supervisors requirements, but is willing to let you try as long as it doesn't distract you from your other duties. The current GAF form may be found on the following two pages. Revise the form according to the changes you believe it needs.

## Follow-up Task 1

After you revise the form, Tonia asks you to present the changes you have made at your weekly staff meeting. Because of a number of other issues on the agenda that day, George Jackson will attend your meeting. Prepare a brief oral presentation that outlines the need for change and the revisions you have offered with your rationale. The presentation should take between 5 and 10 minutes.

## Follow-up Task 2

Tonia has asked you to look into how you would go about storing applications online as well as maintaining records for received applications, recieved payments, participation checks in the Community Work Experience Program, and "proof of use" documents. She is interested in what sort of software could help with this record keeping and whether this will enhance the overall productivity in the office. It has become clear to you that Tonia knows little about her computer and the options it offers beyond word processing and spreadsheets.

Research record-keeping software options and briefly outline your opinion of what may help the office most. Communicate your findings to Tonia in a brief report in memo format.

## Case Appendix: A County GAF Form

Truman County
General Assistance Fund Application

Name ______________________ Phone Number ______________

_____________________________________________
Current Street Address or Rural Route City/Town

How long have you lived at this address? _____
Do you rent or own?_____
If less than a year, what is your previous address?

Address:______________________ City/State ______________

Have you ever been homeless? y/n
If so when and for how long? ______________________

**ALL General Assistance granted must be repaid to Truman County by actual reimbursement to Truman County or by participation in the Community Work Experience Program.**

**Do you agree to repay the amount loaned IN FULL? _____**
**Are you willing to work for your assistance? _____**
**Are you willing to enter into a repayment agreement? _____**

Are you a citizen of the U.S. (or a legal alien?) Yes/No
Are you or is any member of your household a veteran? Yes/No

Individually list EVERY person who lives in the same house as you. Also, list ALL monthly income for each person and the source of that income.

| NAME | BIRTHDATE | SOCIAL SECURITY # | SOURCE OF INCOME | AMOUNT |
|---|---|---|---|---|
| | | | | |
| | | | | |
| | | | | |
| | | | | |

Do you or does anyone who lives with you have any of the following?

| | Y | N | VALUE | | Y | N | VALUE |
|---|---|---|---|---|---|---|---|
| CASH on hand | __ | __ | _____ | Property | __ | __ | _____ |
| Checking account | __ | __ | _____ | Trust Fund | __ | __ | _____ |
| Savings account | __ | __ | _____ | Burial Lots | __ | __ | _____ |
| Vehicles | __ | __ | _____ | Other | __ | __ | _____ |

List Year and Make

______________________

List the FULL and ACCURATE amount spent monthly for the following:

Rent/House Payment $__________ Including Utilities? Yes/No
Which Utilities? __________

Electricity $__________

Gas $__________ Natural or Propane (circle)?

Water/Sewer $__________ monthly or quarterly (circle)?

Medical $__________

Do you have health insurance? Yes No Type____________________

What type and amount of assistance are you requesting at this time?

________________________________________

________________________________________

Have you received General Assistance this year? yes/no Date ______

**IT IS VERY IMPORTANT THAT YOU READ AND UNDERSTAND THE FOLLOWING CERTIFICATION STATEMENT**

**I assume FULL responsibility for the accuracy of the statements made on this application and for any statement made during the application process. I understand that these statements will be used in determining my eligibility for General Assistance and may be investigated and verified.**

**If, for any reason, the previous statements are determined to be false, I understand I may be prosecuted and/or suspended from future General Assistance.**

**I understand that all General Assistance granted MUST be repaid in FULL.**

**I understand that I MUST notify the General Assistance Director of any change in my financial or living situation.**

______________________________ __________

Signature Date

# Summary of Evaluative Criteria for Major Tasks in Case 7

| | ***1 Unacceptable***<br>Insufficient answer to assignment expectations | ***2 Below Average***<br>Inappropriate or ineffective verbal/ visual choices limit document success | ***3 Meets Task Expectations***<br>Has answered objectives of assignment, but individual components could be strengthened | ***4 Above Average***<br>Few flaws, document meets expectations, but could benefit from more attention to detail | ***5 Excellent/ Professional***<br>Few or no flaws, demonstrates keen insight into case subtleties and details |
|---|---|---|---|---|---|
| **Purpose/Key Points**<br>▪ Identifies and meets purpose<br>▪ Articulates key points clearly and with attention to office dynamics | | | | | |
| **Context**<br>▪ Identifies/defines context and situational constraints<br>▪ Demonstrates awareness of GAF form situatedness<br>▪ Revised form demonstrates awareness of opportunities for further development and options | | | | | |
| **Audience**<br>▪ Identifies/defines audience and meets identifiable needs<br>▪ Establishes appropriate tone<br>▪ Understands technical details enough to communicate effectively | | | | | |
| **Design**<br>▪ Demonstrates awareness of visual design elements of task<br>▪ Demonstrates aan awareness of design options & technological aids | | | | | |

# CASE 8

# *Visualizing Technical Information: Designing Agri*Point's Hybrid Production Timetable*[1]

This case emphasizes the importance of design considerations as technical communicators face presenting a variety of information to diverse audiences. Individually or as teams, you are asked to create a variety of visual options for outlining a new hybrid production timetable for a national seed company. Design choices are limited by various print and contextual constraints.

## Background

This paragraph introduces you to Agri*Point. Why is it sometimes useful to define an organization by the size and scope of its competitors?

Agri*Point is a hybrid seed producer based in Kearney, Nebraska. While the home office is located in Kearney, Agri*Point has offices, plants, and labs in many locations in the United States and Canada. The U.S. seed production operations alone include fourteen separate locations, primarily in the Midwest.

| | |
|---|---|
| Addison, IL | Hopkins, MN |
| Flora, IL | Oakdale, MN |
| Algona, IA | Blue Springs, MO |
| Grinnell, IA | Sioux Falls, SD |
| Alta, IA | Watertown, SD |
| Kearney, NE | Callahan, TX |
| De Witt, TX | Junction City, KS |

An employer of approximately 9,500 people, Agri*Point is certainly not as large as the major producers and research organizations like Ciba, DeKalb, or Pioneer Hi-Bred International, but the company is widely recognized as a strong mid-range seed producer in the United States.

Here you learn Agri*Point's purpose and goals. How does the mission statement help you to understand what is important to Agri*Point?

Agri*Point tests, analyzes, and produces sunflower, corn, and soybean seeds. The mission of the organization is "to develop breeding technologies and appropriate biotechnological research methodologies that address the demands of changing climatological and agronomic factors and meet the needs of today's

[1] I wish to thank Rod Muilenberg for his expertise in helping develop the technical components of the background for this case.

farmers." Like most producers, Agri*Point works to develop the perfect hybrid seeds that adapt to a variety of soil and weather conditions. The company serves only the United States and Canada; however, it has recently discussed expansion into New Zealand and Thailand.

Agri*Point's hybrid production is a complex process of development and analysis that spans several years before hybrid seeds are ready to be marketed. The testing for viability of a new hybrid involves careful analysis of several important characteristics of the sunflower, corn, and soybean seeds. Agri*Point employs the internationally accepted ISTA seed testing methods. Specifically, Agri*Point focuses on the following:

- *Germination* (the length of time for the plant to develop from the seed in normal field conditions): Replicates (hybrid test subjects) of seeds are extracted from the pure seed fraction of the sample and germinated in controlled conditions. Agronomists then evaluate and classify the viability of the seedlings.
- *Vigor* (the potential for development): Seeds undergo stress tests to determine potential for development under varying conditions.
- *Purity analysis* (seeks the number and weight of contaminants in a specific quantity): Tests specific quantities of subjects for infiltration of outside contaminants such as noxious weeds.
- *General seed health* (seeks the presence of specific fungi or diseases known to affect seed germination and growth): Aims to grow fungi or disease and test seed resistance. Some specific diseases include (but are not limited to): Blackleg, Alternaria, Sclerotia, Smut, Fusarium, Anthracnose.

Maturity rate (or the RM count) is just one technical characteristic important to seed producers and consumers. Highlight as many other characteristics that appear important in judging the marketability of seed corn.

While development of new hybrid seeds often takes several years, so too does the testing in the postdevelopment stage prior to marketing. One characteristic routinely tested is relative maturity. The relative maturity rate indicates how long a seed will take to mature to harvest, and most consumers purchase seeds based on how the maturity rate fits the agronomic needs of the area. For example, in the Northern Plains states and several Canadian farm areas, farmers require a shorter relative maturity rate for hybrid corn seeds (e.g., 99 days as opposed to the longer 115–120 day RM), because spring planting weather comes later, often by as much as two weeks, and first frost comes earlier than in the more central or southern regions. For corn hybrids, one component of judging relative maturity is to test the moisture content of the kernels at or near the target relative maturity date. Corn must dry prior to harvest for storage purposes; thus, the drying rate is essential to determining maturity.

Most U.S. states require at least three years of field testing in order to determine the relative maturity rate. Because the climatological variances determine the appropriateness of a given hybrid in a specific region, states often stipulate relative maturity testing and product marketing limitations. This is, in part, to prevent marketing a hybrid at a faster or slower relative maturity than it actually is in order to bolster sales in that region. In most cases, states also require annual registration of each hybrid seed used or tested to monitor production and soil effects.

The main point is that developing, testing, and marketing new hybrids takes time and is sometimes dependent on unstable factors such as the weather.

Therefore, because development often takes several years, field testing takes several more, and compliance with governmental regulations also slows the path to market, hybrid seed production is an ongoing and time-consuming process. Agri*Point is no different from most other seed production companies, then, in its need for careful development planning and communication of these plans to all components of the team including agronomists, field testing agents, and marketers.

## The Situation and Your Role

Your role is defined here. What skills or training do you believe a field coordinator might need?

You are a field coordinator for corn hybrid seed production in the regional Agri*Point office in Sioux Falls, South Dakota. As a field coordinator, you help to track development and production of new hybrids, and set up test plot caretakers. Test plots are basically square acres of land upon which a farmer or co-op agrees to plant and test a specific hybrid and eventually share results with the company and consumers. Often, farmers designate several acres of land to test plots for a specific company. Then, close to harvest, the company hosts a Field Day for hybrids that are ready for market. Sales representatives and prospective customers then attend the Field Day to inspect the preharvest yield.

As a field coordinator, you normally receive information from agronomists and developers about which hybrids are coming available and the timeframe in which you can expect to usher them into field testing. This communication usually comes in the form of an internal report from the main office in Kearney, where the bulk of the genetic testing is initiated. The report is usually issued on or around January 1 of each year so that field coordinators can contact test sites and managers enough in advance of planting season. In your region, you have coordinated field testing for anywhere between two to six new hybrids in a given season. After the test plots are established, you work with a team of specialists to maintain daily growth charts and disease test results. You then summarize the season's data and report it to the marketers so that they may use it to sell the product when the seeds are ready for consumer distribution.

This paragraph indicates a change in normal procedure for the company.

This year, however, your supervisor, Malcolm Sands, has explained to you that Agri*Point upper management has requested that all regions offer projections for the development and testing schedules for the years 1999 through 2004. In compliance with new federal laws, all seed producers must anticipate market entry by as much as three years.

This paragraph explains the limits the company has imposed on the field coordinator's approach to the task. Why would Agri*Point's main office impose such restric-

Agri*Point plans to create a booklet highlighting each region's production plans through 2004. Each region is alloted only limited space to communicate the data and your region is allotted only 6 x 6 inches on a standard white sheet of paper to illustrate your projections. You have been limited not only by space but also by the fact that you will not be allowed to use color. In addition, some tions? How do these limits affect the approach you take to communicating this information?

of the information, because it is based on projections, is incomplete. You may use charts, graphs, tables, text, or some combination of any option to communicate your projections. The summary of projections is outlined in the case appendix.

## Background Development

Consider the following prompts as a way to evaluate what you've read in this case. Answer the questions on paper or with others online, or work in a group and discuss possible answers orally. You may use your answers to help you in subsequent tasks.

- Agriculture is an essential industry to the American economy. As in other industries, agricultural leaders constantly seek to improve the products' technology. This case deals with genetic testing and development of new products. In a small group or with a partner, search the Internet and make a list of five other ways that agricultural industry has sought to make changes or improvements over the past decade. Discuss your findings with the class.
- Highlight and look up the meanings of any technical terms in the case that you may not understand. Together with your classmates, create a glossary of terms for your reference.
- Based on the description of the duties of a field coordinator, what specific skills or training do you believe an individual might need to be successful in such a role? Write out your answers and share them with the class.
- What are the factors that limit your options for how best to communicate the requested information? If you had no limits, how might you answer the need differently?

## Gathering More Information

Dekalb and Pioneer Hi-Bred International, Inc. both have websites that will be useful to you in understanding agronomic strategies, genetic testing, and hybrid seed marketing. With a partner, access the sites **http://www.pioneer.com** and **http://www.dekalb.com**. Access the products pages to gain insight into the kinds of genetic testing and development the major seed producers do. These pages should also offer you some design ideas for summarizing large amounts of information succinctly.

## Major Task 1

Using the information provided in the case appendix, design two different options of how to condense and communicate the projections to a 6 x 6-inch space on paper. The design choices you make should aim to clearly and accurately communicate the information you have to internal readers (upper management at Agri*Point) as well as to external readers (U.S. Department of Agriculture readers).

## Follow-up Task

Write a memo to your supervisor, Malcolm Sands, that serves as a cover letter to the designs you've offered, explaining your choices and your preference for publication in the booklet.

## Major Task 2

Agri*Point has decided to incorporate the development projections on a website dedicated to corn hybrids. Redesign your information (with no spatial or color constraints) to fit the needs of a broader and less specialized readership on the Internet. You may want to discuss with class members how the rhetorical demands of the two publications require different design strategies.

# Case Appendix: Designing Agri*Point's Hybrid Production Timetable

The following summary represents the list offered by Agronomist Todd Honold and his team for likely predictions of hybrid seed production, testing, and target market dates. This list was accompanied by a memo in which Honold is adamant about offering some sort of disclaimer to projections. He noted, "These are our best guesses. But I emphasize that in any given year, we can face problems in testing, climatalogical anomolies, patent hold-ups, and any number of other stumbling blocks to production. Any one of the hybrids we're listing here runs the risk of dropping off the production list at any time. I understand that the request is for projections, but I think it's very important to explain that the list is not fixed or stable somewhere in your report."

## Projections

### Hybrids

**AP-2890**
This hybrid is currently in testing and scheduled for market for the 1999 season. It is in the early/medium maturity range at 97RM. Based on preliminary testing, drying and emergence ratings look good to excellent. Best for north-central region soils and tested in Mankato, Minnesota and Ellis, North Dakota.

**AP-2920**
This hybrid is currently in development and projections for testing are for 2000. It is likely, with optimal conditions, AP-2920 will be ready for market by 2002 or 2003. It is in the early/medium maturity range at 100RM. Unable to judge drying and emergence ratings yet. Also best for north-central region plot testing.

**AP-2850**
This hybrid is currently in testing and scheduled for market release in the 1999 season. It is in the early/medium maturity range at 99RM. Based on preliminary testing, drying and emergence ratings look excellent, as does disease resistance. Versatile and hearty. Will do well in north, north-central, west-central, and east-central regions. Plot tested in Rainy Lake, Ontario, Canada; Sloan, Iowa; Dale, Minnesota; and Westburg, North Dakota.

**AP-3100**
This hybrid is currently in development and projections for testing are for 2000. It is likely, with optimal conditions, AP-3100 will be ready for market in 2002 or 2003. It is in the medium maturity range at 107RM. Unable to test drying or emergence ratings yet, but the seed has been engineered for especially fast drying. Best for central regions (southern Iowa, Missouri, and Kansas).

**AP-3450**
This hybrid was recently proposed by a team at Kearney studying high oil and nutritionally enhanced seeds. The team hopes to patent a gene combination associated with this hybrid. Because it is still in the early development stage, projections for testing are questionable. If a patenting process ensues, the development and testing may be accelerated. Currently projections indicate testing for 2001 and market readiness for 2003 or 2004. It will be in the medium maturity range at 106-

108RM. Unable to speculate on drying or emergence ratings yet. Best for central regions (southern Iowa, Missouri, Kansas).

**AGRI-2040**
This hybrid is currently in development and projections for testing are for 2001 with possible market release in 2002 or 2003. Will be in medium to late maturity range at 115RM. Designed for drier conditions, this hybrid is especially drought resistant. Unable to test for drying or emergence ratings. Recommended plot testing in Texas.

**AGRI-2060**
This hybrid is currently in testing and projections for possible market release are in 2000 or 2001. It is in the medium to late maturity range at 114RM. Based on preliminary testing, drying ratings look good and emergence ratings look excellent. Stalk and root strength are average to good. Will do well in west-central, east-central, and south regions. Plot tested in Dale, Texas; Hannibal, Missouri; and Farnsworth, Oklahoma.

**AGRI-2080**
This hybrid is nearing readiness for market and should be released in 1999. It is in the medium maturity range at 106RM. Testing indicates that drying ratings look good to excellent. Emergence looks average, and stalk and root strength ratings are good to excellent. Successfully plot tested in Flora, Illinois; Kearney, Nebraska; Ottumwa, Iowa; and Newton, Missouri.

**EB-1820**
This hybrid is currently in development and projections for testing are for 2000. Under optimal conditions, this seed may be released for market sometime after 2002. This hybrid is designed for a rapid maturity rate at 90RM and will offer nutritional enhancements. Emergence and drying rates are untested at this time. EB-1820 will likely be tested in Iowa, Nebraska, and South Dakota.

**EB 1850**
This hybrid is currently in development and projections for testing are for 2000. Under optimal conditions, this seed may be released for market sometime after 2002. This hybrid is designed for rapid maturity rate at 94RM and is engineered for sweeter taste. Emergence and drying rates are untested at this time. EB-1850 will likely be tested in Iowa, Nebraska, and South Dakota.

## Specialty Hybrid Seeds

**ASP-2020B**
This hybrid is currently in development and projections for testing are for 1999. Under optimal conditions, this seed may be released for market in 2000 or 2001. The patent is pending for this Asian-corn borer resistant seed, genetically engineered with a combination of insect-tolerant genes. This hybrid is designed for early to medium maturity rate at 100RM. Emergence and drying rates are anticipated to be excellent. Because the new Asian-corn borer problems have been most prevalent in the southern regions, focus on testing will take place largely in region 8. Plot testing is scheduled for McCook, Nebraska; DeWitt, Texas; Brunswick, Georgia; and Ada, Oklahoma.

**ASP-2250**
This hybrid is currently in development and projections for testing are sometime after 2000. Under optimal conditions, this seed may be released for market by 2003 or 2004. Development is focusing on engineering a pendimethalin herbicide-resistant seed. This hybrid is designed for medium maturity rate at 106RM, and emergence and drying rates are expected to be excellent. It will be tested in Junction City, Kansas; Blue Springs, Missouri; and Keokuk, Iowa.

**ASP-2400**

This hybrid is currently in development and projections for testing are for 1999. Under optimal conditions, this seed may be released for market by 2001. A patent is pending for this Black Leaf Spot resistant seed. Black Leaf Spot, which emerged in Europe in 1995, caused serious damage to crops in Iowa and Nebraska in 1997. This hybrid is designed for early to medium maturity rate at 99RM, and emergence and drying rates look good to excellent. Plot testing is scheduled for Grand Junction, Iowa; Greenfield, Nebraska; and Hull, Iowa.

**ASP-0020**

This new hybrid will be ready for market in 1999. ASP-0020 is a popcorn seed designed for early to medium maturity rate at 100 (normal popcorn RM is 110, so this will be attractive to especially eastern regions that risk later planting dates). Projections suggest an upward trend in popcorn production over the next five years. Plot testing in Sac City, Iowa and Flora, Illinois indicates emergence and drying rates are good to excellent. Roots and stalk strength are average to good.

**ASP-1500**

This hybrid is currently in development and projections for testing are scheduled for 2000. Under optimal conditions, this hybrid will be released for market in 2001 or 2002. This sweet corn is genetically engineered to resist bromoxynil-based herbicides and is projected to have a medium maturity rate at 108RM. Testing has not yet offered projections of drying or emergence.

# Summary of Evaluative Criteria for Major Tasks in Case 8

| | *1 Unacceptable*<br>Insufficient answer to assignment expectations | *2 Below Average*<br>Inappropriate or ineffective verbal/ visual choices limit document success | *3 Meets Task Expectations*<br>Has answered objectives of assignment, but individual components could be strengthened | *4 Above Average*<br>Few flaws, document meets expectations, but could benefit from more attention to detail | *5 Excellent/ Professional*<br>Few or no flaws, demonstrates keen insight into case subtleties and details |
|---|---|---|---|---|---|
| **Purpose/Key Points**<br>▪ Identifies and meets purpose<br>▪ Articulates key points clearly<br>▪ Summarizes key issues with attention to clarity and creativity | | | | | |
| **Context**<br>▪ Identifies/defines context and situational constraints<br>▪ Demonstrates awareness of project's limits | | | | | |
| **Audience**<br>▪ Identifies/defines audience and airms to meet identifiable needs<br>▪ Establishes appropriate tone<br>▪ Understands technical details enough to communicate effectively | | | | | |
| **Design**<br>▪ Demonstrates awareness of visual design elements of task<br>▪ Demonstrates an awareness of design options & technological aids in the development of these options<br>▪ Demonstrates awareness of differences between print and electronic design considerations | | | | | |

CASE 9

# Charting Immigration Changes: Communicating Visually with Statistical Data

The task is to visually "translate" technical data in several ways to determine the most effective means of communicating statistical trends for the U.S. Department of Immigration and Naturalization. You are challenged to make an argument with the data as well as face conflicting purposes articulated by managers.

## Background

Understanding the history of immigration is important for a comprehensive examination of possible trends. Do a library search of immigration sources and familiarize yourself with the historical changes.

Immigration laws and the effects of illegal immigration on U.S. communities have recently been the focus of several hot political debates. In California—the state arguably affected by illegal immigration more than any other—this issue has provided such intense political discussion at state and local levels that it thrust the issue into the national spotlight. The political hot button of immigration, however, certainly predates the recent debates in California and elsewhere.

Since the 1880s, American immigration policy has worked to limit and control the increasing trend of people seeking to locate in the United States. The historic Foran Act in 1885, for example, prohibited businesses and individuals from recruiting unskilled labor from outside the U.S. with advance contracts. From 1890 to the mid-1920s, new literacy requirements further restricted immigration. In fact, the first Immigration Act in 1917 required literacy for immigrants over the age of 16. Many sociologists and political scientists contend that the restrictionist policies associated with immigration since the late 1800s have largely emerged from socially and politically sanctioned racism and a protectionist attitude among native-born American citizens.

Two important recent pieces of legislation are relevant here. The first is the Immigration Reform and Control Act of 1986. Signed into law by Ronald Reagan in November 1986, the bill contains the following noteworthy provisions:

Why is it helpful to understand the specifics of legislative changes?

- Employers would be forbidden to hire illegal aliens. The ban would apply to all employers, even those with just a few employees.
- Employers would have to ask all job applicants for documents, such as a passport or birth certificate and a driver's license, to confirm that they were

either citizens or aliens authorized to work in the United States. The employer would not be required to check the authenticity of the document.

- The government would offer legal status to aliens who entered the United States illegally before January 1, 1982 and have resided here continuously since then. For five years, they would be ineligible for welfare, food stamps, and most other federal benefits, with some exceptions.
- States would have to verify, through records of the Immigration and Naturalization Service, the legal status of aliens seeking welfare benefits, Medicaid, unemployment compensation, food stamps, housing assistance, or college aid under federal programs.
- Under a special program, illegal aliens who worked in American agriculture for at least 90 days in the period from May 1, 1985, to May 1, 1986, could become lawful temporary residents in the United States. After two years of that status, they could become permanent residents, eligible for American citizenship after five more years.
- Employers would be forbidden to discriminate against legal aliens because of their national origin or citizenship status. A new office would be established in the Justice Department to investigate complaints of such discrimination.
- To improve enforcement, the Immigration and Naturalization Service would receive $422 million more in fiscal year 1987 and $419 million in fiscal year 1988. The agency's budget in FY 1987 was $593.8 million, of which $379.7 million was for enforcement.[1]

The second bill, the Immigration Act of 1990, was signed into law by President George Bush in November of that year. The new law placed tighter restrictions on the number of immigrants allowed to enter the U.S. annually. According to the act, the number of visas granted to immigrants (excluding refugees) would drop from the 1992–94 figure of 700,000 annually to 675,000 beginning in fiscal year 1995. The history of American political attention to immigration issues has varied widely across decades since the late 1800s; however, the legislative trends have leaned toward tighter restrictions and attention to economic and social effects of immigrant populations on American communities.

## The Situation and Your Role

Your role is defined here. What do you anticipate you would do in this role?

You were hired just yesterday as a researcher and speechwriter for Brian Lindeman, candidate for U.S. Congress from Worscester, Massachusetts. Lindeman is running as an independent. New on the political scene, he has spent most of his career in higher education teaching political science and ethics. He

[1] Francesco Cordasco, *Dictionary of American Immigration History* (Metuchen, NJ: Scarecrow Press, 1990), xxii.

cultivates his Washington "outsider" status and is just in the process of getting his campaign (and fundraising) off the ground. Though you have yet to meet Lindeman personally—you were hired by his chief of staff, Shaswana Bahktu—you feel strongly that this professional opportunity is a good chance to keep your professional communication skills polished as well as to enhance your portfolio. You have read most of what has been written about Lindeman (which is not much) and feel there is nothing in what Lindeman has articulated as his position on various issues with which you vehemently disagree, although your interest in politics is certainly secondary to your interest in professional communication.

On your first day of work, you are anxious to get started on a project and Bahktu, your immediate supervisor, does not disappoint you.

"We're trying to get a handle on several issues that the media have targeted as most important in this election," she explains as you sit down with her at her cluttered desk. "I think I'm going to get you started out with some research and see what it tells us."

You're a little disappointed because you hoped you'd start right away with some writing, but you nod. "What are we investigating?"

Immigration has been a hot topic in California and New York. Many of California's teachers have refused to comply with recent law changes regarding illegal aliens in the school systems there.

"We're starting out with immigration," she responds, handing you several sheets of paper with tables and figures. "Although immigration has been hot mostly in California and New York, this state, and especially this city, is not immune to the issue at all. What we need is to collect some data and sift through it to see specifically what it tells us in terms of trends."

"For Worcester? The state? The nation? What kind of trends are we looking for?" You quickly glance at the pages she has handed you.

A trend is a pattern of change that can be observed over a sustained period of time.

"Well, we're starting sort of from ground zero here, so I can't really tell you what sort of trends we're looking for. The information we come up with will need to be organized and clearly articulated so that Brian can take a public stance on the issue if asked. We're interested in what the immigration statistics can tell us about changes in the U.S. population, the implications—mostly economic and social effects—of illegal immigration, and any special focus you can find on this area, obviously, since we're working to sell what we say to the voters *here*."

Your purpose is defined here. Is your initial understanding of what you are to do compatible with your superior's?

You believe your assignment is really no different than creating a research database, so you feel confident that this is something you can accomplish. Shaswana Bahktu points to the pages she's handed you, "Those are some preliminary statistics I dug up from some common sources, but you'll likely want to find more. Then I think it would be good to briefly outline whatever arguments you feel like you can make, given the numbers you've uncovered. Obviously, not all of the numbers are going to be relevant to what Brian says about this issue, so you need to make this as accessible as possible for him. Tables are sometimes sort of hard to read, so figure out how to make your point *quickly*, so that I can glance at this and know what you're trying to show. You can use whatever means feel right depending on the data and what it is you're trying to show. I'd like you to organize what you come up with and put it on my computer by the end of the day tomorrow—is that enough time?" She glances at her watch and begins searching her desk for something.

"Tomorrow? Well, sure," you answer hesitantly. "Um ..."

She grabs a stack of folders and throws a purse over her shoulder. "I've got to be out of the office most of today, so if you have any questions, ask Greg Fait. He's doing something similar with abortion and has been with us for several months." She gestures to a bookcase with several large volumes in it. "You can use our resources for research, head to the library, or make calls. I don't really care how you get the info; just make sure it's right."

With that, the young woman quickly leaves the office, and you make your way back to your new desk. You look at the statistics she has provided you. They're a start, but you know you'll need more than you have. Just as you begin to write some notes about the data you have, Greg Fait wanders over to your desk.

He holds out his hand to shake. "Nice to meet you," he says. "So you and I are working similar projects, eh?"

You nod. "I'm working on immigration."

What is the challenge that Fait identifies here?

Greg shakes his head. "That's a toughie. Lindeman hasn't made up his mind where he stands on that one, so you're probably starting from scratch, aren't you? At least with abortion, he had an ideological stance figured out. My job has been easy enough just because I know what position to look for in the data. Need any help?"

"Well, I guess I need to collect some statistics first, but it might be good to know what Shaswana is looking for in my report tomorrow. She didn't really have time to tell me."

"She only gave you until tomorrow?" Greg laughs and shakes his head. "It's the classic test. See how much the new kid can come up with in the shortest amount of time. Well, I can tell you that she's going to look for you to understand what the trends in the statistics are saying. If I were you, I'd carve out three or four good arguments that you can back up with the numbers and have those outlined for her. You know, like welfare benefits in X community have risen proportionately with the rise in suspected illegal immigration, something like that."

One of the problems with researching an issue like immigration is that trends in the data can be reduced to stereotypes. Be careful to avoid the trap of simplifying data too much.

"Wait!" You hold up your hand to stop him as you're taking notes. "I don't want to perpetuate stereotypes here. Are you suggesting that the arguments I'm making are going to be *advice* to Lindeman for his stance?"

"Not directly, but the research you do is the foundation on which he bases whatever stance he takes. My example was just off the cuff, really."

You shake your head. "But if these arguments are the foundational research, how can Shaswana expect me to have this to her tomorrow?"

Greg laughs. "Aw, don't worry about it. She's not really expecting spit and polish or even a complete report. She wants to see what you can do in the time she's given you. It's indicative of what we'll face once things come down to the wire around here. Anyway, I'd say she's probably looking for evidence that you can write a good text summary. Don't mess around with a huge amount of detail."

After a few more minutes of chatting, you and Greg talk about getting a pizza after work. When he leaves, you return to your notes. You are just getting ready to head to the library when Connie Reagan introduces herself to you as one of the head writers for Lindeman. She seems interested in your background and wants to chat, but you are feeling the pressure of time to get at the research as quickly as you can. After about 15 minutes of discussion about backgrounds and interests, you explain to Connie that you need to get started on your project for Shaswana. Connie asks you about what you are researching.

"I'm doing some base research on immigration."

"Ah," she nods. "You're on base detail, huh? Want some free advice?" she asks, smiling.

"Yes, of course. I can use any help I can get," you answer, hoping she'll give you the same insight into Shaswana's preferences that Greg provided earlier.

How does Reagan's understanding of Shaswana Bahktu's expectations differ from Fait's?

"Present your information as economically as you can and focus as much as possible on graphics and visuals to make your point. Also, Shaswana's a stickler for detail and completeness, so don't bite off more than you can chew. When did she ask you to get it to her?"

Your heart sinks a little at the apparent contradiction you've just heard between Greg and Connie. "Late tomorrow."

Connie smiles. "Hope you weren't planning on getting any sleep tonight."

## Background Development

Consider the following exercises as a way to fine-tune your understanding of the case and its details. Answer the prompts alone or work in small groups of two or three to discuss the answers. Feel free to draw on your responses to inform any of the tasks that follow.

- This case has asked you to research and assimilate a great deal of statistical data in order to form an argumentative stance on immigration issues. Such a challenge requires that you do background research, particularly since Shaswana has given you only some bare bones material. Form a group with at least two other peers from your class. Assign each individual in the group a research domain. For example, one group member may be in charge of scanning the Internet or accessing the U.S. Census Bureau's World Wide Web site. Another person might be in charge of searching through statistical abstracts or census reports. When all group members have completed some preliminary research, meet as a group and exchange the information you have amassed. Scan the information for trends. As a group, record at least five assumptions you feel you can make based on the data you have uncovered.

- Because you must ultimately narrow your field of inquiry with this data, you must begin with some questions you are interested in answering with the data. Write down three questions pertaining to immigration that you would like to see if you can answer with the data. When you have written the questions, discuss and compare them with others in your class. Help each other to hone the questions so that they may be answered by the data.

- Statistical data are sometimes difficult to talk about and even understand unless they are offered in a way that summarizes a primary argument. For example, if you are interested in knowing what the trends are in arrests made at three strategic border crossings over the past ten years, you will likely have to sift through a great deal of statistical data in order to get to your answer. As a result, making data accessible is often of chief concern to technical communicators. Choose one of the trends you wish to highlight and think about how to simplify the information visually. Consult your text on visual strategies you may employ. Experiment with two or three different visual appeals (e.g., charts, graphs, lists). Then share your experiments in visual communication with others in a small group. Examine each others' work for clarity, reading ease, and visual appeal.

## Gathering Information

With the statistics Shaswana has provided you in the case appendix and the statistics you uncover through your own research, write out a list of five questions you would like the data to answer and possible corresponding trends you feel are supported by the numbers. These are only to serve as personal notes, so you may use this task as a prewriting journal, if you wish.

## Prewriting Task

Greg's and Connie's appraisals of what Shaswana is after for your report seem on the surface to conflict with one another. Make an outline for your report using your best judgment about what Shaswana is after and what would be most beneficial for the candidate as he creates a public statement. Again, this outline may be a part of your prewriting journal.

## Major Task 1

You have decided to use a balanced approach to the final report Shaswana has assigned you, so you have determined to use approximately half visuals and half text. Design three or four visuals you believe would be most accessible and appropriate for communicating the trends from the statistical data you have uncovered and write complementary text summaries with them.

## Major Task 2

Finalize your report (in memo form to Shaswana Bahktu) and be prepared to explain the rationale you used to highlight the materials.

## Follow-up Task

Shaswana Bahktu appreciated your initiative on the statistical analysis and report. She has asked you to think of an overriding slogan, a sound bite, that the candidate might use to summarize his stance on immigration issues. How might you summarize what your statistics suggest in a campaign slogan? In a brief memo to Shaswana, offer one or two suggestions for slogans and why you think they would be effective. Focus on the slogans' relationship to your statistical data and make brief arguments for why you believe they are effective.

# Case Appendix: Immigration Tables

## Immigrants Admitted, by Type and Selected Class of Admission[1]

| Type and Class of Admission | Fiscal Year 1991 | Fiscal year 1992 | Fiscal Year 1993 |
|---|---|---|---|
| **Total, all immigrants** | **1,827,167** | **973,977** | **904,292** |
| New arrivals | 443,107 | 511,769 | 536,294 |
| Adjustments | 1,384,060 | 462,208 | 367,998 |
| **Total, IRCA legislation** | **1,123,162** | **163,342** | **24,278** |
| Residents since '82 | 214,003 | 46,962 | 18,717 |
| Special (agriculture) | 909,159 | 116,380 | 5,561 |
| **Total, nonlegalization** | **704,005** | **810,635** | **880,014** |
| **Preference immigrants** | **275,613** | **329,321** | **373,788** |
| **Family sponsored** | **216,088** | **213,123** | **226,776** |
| Unmarried children | 15,383 | 12,486 | 12,819 |
| Spouses of U.S. citizens | 110,126 | 118,247 | 128,308 |
| Married children | 27,115 | 22,195 | 23,385 |
| Siblings of U.S. citizens | 62,462 | 60,195 | 62,264 |
| **Employment-based immigrants** | **59,525** | **116,198** | **147,012** |
| Priority workers | NA | 5,456 | 21,114 |
| Professionals | NA | 58,401 | 29,468 |
| Skilled workers | NA | 47,568 | 87,689 |
| Special immigrants | 4,576 | 4,063 | 8,158 |
| Employment creation | NA | 59 | 583 |
| Pre-1992 | 54,949 | 651 | NA |
| **Immediate relatives of U.S. citizens** | **237,103** | **235,484** | **255,059** |
| Spouses | 125,397 | 128,396 | 145,843 |
| Children | 48,130 | 42,324 | 46,788 |
| Orphans | 9,008 | 6,536 | 7,348 |
| Parents | 63,576 | 64,764 | 62,428 |
| **Refugees and asylees** | **139,079** | **117,037** | **127,343** |
| Refugee adjustments | 116,415 | 106,379 | 115,539 |
| Asylee adjustments | 22,664 | 10,658 | 11,804 |
| **Other immigrants** | **52,210** | **128,793** | **123,824** |

[1] Data from Immigration and Naturalization Service, U.S. Department of Justice, reprinted in *World Almanac Book of Facts 1995* (New York: Funk & Wagnalls, 1995) p. 385.

## Immigrants Admitted for Top 30 Metropolitan Areas of Intended Residence (1993)[1]

| Metropolitan Statistical Area of Intended Residence | Total Number | Total Percentage | Metropolitan Statistical Area of Intended Residence | Total Number | Total Percentage |
|---|---|---|---|---|---|
| **Total** | **904,292** | **100.00** | 16. Nassau-Suffolk, NY | 11,601 | 1.3 |
| 1. New York, NY | 128,434 | 14.2 | 17. Seattle, WA | 11,509 | 1.3 |
| 2. Los Angeles, CA | 106,703 | 11.8 | 18. Riverside, CA | 11,187 | 1.2 |
| 3. Chicago, IL | 44,121 | 4.9 | 19. Dallas, TX | 10,959 | 1.2 |
| 4. Miami-Hialeah, FL | 30,464 | 3.4 | 20. Detroit, MI | 9,816 | 1.1 |
| 5. Washington, DC | 27,427 | 3.0 | 21. Jersey City, NJ | 8,754 | 1.0 |
| 6. Orange County, CA | 24,921 | 2.8 | 22. Ft. Lauderdale, FL | 8,124 | 0.9 |
| 7. Houston, TX | 22,634 | 2.5 | 23. Atlanta, GA | 8,031 | 0.9 |
| 8. San Francisco, CA | 21,054 | 2.3 | 24. Middlesex, NJ | 7,371 | 0.8 |
| 9. Boston, MA | 20,414 | 2.3 | 25. Honolulu, HI | 6,880 | 0.7 |
| 10. San Jose, CA | 19,473 | 2.2 | 26. Fresno, CA | 6,780 | 0.7 |
| 11. San Diego, CA | 16,931 | 1.9 | 27. Sacramento, CA | 6,746 | 0.7 |
| 12. Oakland, CA | 16,087 | 1.8 | 28. Mpls-St. Paul, MN | 6,349 | 0.7 |
| 13. Newark, NJ | 13,551 | 1.5 | 29. Phoenix-Mesa, AZ | 5,918 | 0.7 |
| 14. Bergen-Passaic, NJ | 12,931 | 1.4 | 30. Portland, OR | 5,800 | 0.6 |
| 15. Philadelphia, PA | 12,842 | 1.4 | | | |

[1] Data from Immigration and Naturalization Service, U.S. Department of Justice, *World Almanac Book of Facts 1995* (New York: Funk & Wagnalls, 1995), p. 385.

## Estimated Undocumented Immigrants by Selected States and Countries of Origin: 1992 and 1994[1]

[**In thousands.** The ranges of estimates supplied by the Bureau of the Census represent indicators of magnitude which the Bureau believes is responsive to the inherent uncertainty in the assumptions underlying these estimates. The Census indicators are the results of applying 3 calculated percent distributions on two national estimates of the undocumented population. The 3 percent distributions by state are based on: (1) an average between the number of undocumented immigrants by state included in the 1980 census and of the number of legalization applicants by state produced by the Immigration Reform and Control Act (IRCA) of 1986; (2) the number of undocumented immigrants by state estimated by the Immigration and Naturalization Service (INS); and (3) the number of foreign born noncitizens by state counted in the 1990 census. The figures supplied by INS are based on estimates of illegal immigrant population who established residence in the United States before 1982 and did not legalize under IRCA and annual estimates of the number of persons who enter surreptitiously across land borders and nonimmigrant overstays who established residence here during the 1982 to 1992 period. The estimates for each country were distributed to states by INS based on U.S. residence pattern of each country's total number of applicants for legalization under IRCA.]

| State | Bureau of the Census Estimates, 1944 | | INS Estimates, October 1992 | Country | INS Estimates, October 1992 |
|---|---|---|---|---|---|
| | Low | High | | | |
| **U.S. Total** | **3,500** | **4,000** | **3,379** | **Total** | **3,379** |
| California | 1,321 | 1,784 | 1,441 | Mexico | 1,321 |
| New York | 462 | 539 | 449 | El Salvador | 327 |
| Texas | 300 | 427 | 357 | Guatemala | 129 |
| Florida | 243 | 385 | 322 | Canada | 97 |
| Illinois | 157 | 225 | 176 | Poland | 91 |
| New Jersey | 98 | 168 | 116 | Philippines | 90 |
| Massachusetts | 42 | 106 | 45 | Haiti | 88 |
| Arizona | 50 | 68 | 57 | Bahamas | 71 |
| Maryland | 29 | 63 | 27 | Nicaragua | 68 |
| Virginia | 37 | 63 | 35 | Italy | 67 |
| Washington | 32 | 59 | 30 | Honduras | 61 |
| Michigan | 10 | 53 | (NA) | Colombia | 59 |
| Pennsylvania | 15 | 51 | 18 | Equador | 45 |
| Connecticut | 13 | 46 | 15 | Jamaica | 42 |
| Georgia | 29 | 36 | 28 | Dominican Rep. | 40 |
| Ohio | 7 | 36 | (NA) | Trinidad/Tobago | 39 |
| Colorado | 22 | 29 | 22 | Ireland | 36 |
| Oregon | 21 | 27 | 20 | Portugal | 31 |
| Hawaii | 4 | 25 | (NA) | Pakistan | 30 |
| New Mexico | 14 | 25 | 19 | India | 28 |

[1] U.S. Bureau of the Census, "Illustrative Ranges of the Distribution of Undocumented Immigrants by State" by Edward W. Fernandez and J. Gregory Robinson, *Technical Working Paper* No. 8, October 1994; and the U.S. Immigration and Naturalization Service, *Statistical Yearbook*, annual; *Statistical Abstract of the United States, 1995*, p. 12.

# Summary of Evaluative Criteria for Major Tasks in Case 9

| | ***1 Unacceptable***<br>Insufficient answer to assignment expectations | ***2 Below Average***<br>Inappropriate or ineffective verbal/ visual choices limit document success | ***3 Meets Task Expectations***<br>Has answered objectives of assignment, but individual components could be strengthened | ***4 Above Average***<br>Few flaws, document meets expectations, but could benefit from more attention to detail | ***5 Excellent/ Professional***<br>Few or no flaws, demonstrates keen insight into case subtleties and details |
|---|---|---|---|---|---|
| **Purpose/Key Points**<br>▪ Identifies and meets purpose<br>▪ Articulates key points clearly<br>▪ Demonstrates research initiative and appropriate understanding of statistical data | | | | | |
| **Context**<br>▪ Identifies/defines context and situational constraints<br>▪ Demonstrates understanding of arguments that employ supportive evidence with statistics | | | | | |
| **Audience**<br>▪ Identifies/defines primary, secondary and tertiary audiences and their needs<br>▪ Establishes appropriate tone<br>▪ Reads and responds to technical information effectively<br>▪ Visual displays are accessible to the layperson | | | | | |
| **Design**<br>▪ Demonstrates awareness of visual design elements of task<br>▪ Demonstrates an ability to design and explain statistical information | | | | | |

# PART III

# *Applying the Communicator's Techniques*

# CASE 10 *Ensuring Safety in a Hazardous Workplace Environment: Revising and Editing Safety Instructions for Quality Electric*

This case emphasizes the importance of locating flaws in technical communication and tact in responding to communication problems. As team members of a quality control effort in a relatively small electrical plant, you are asked to review new safety instructions and processes. In the review, you will discover problems with the design and effectiveness of the instructions and be asked to revise and report on the changes you have made.

## Background

It is sometimes useful to learn a little background information on the subjects of your cases. Do some preliminary research into hydroelectric energy.

Redbone Falls is a small Canadian village in the province of Ontario with little claim to fame beyond its hydroelectric power plant. Finished and brought on line in 1995 through numerous grants, the Canadian Power Authority, and with the support of Christopher Carter and his team from Magi Engineering, the hydroelectric plant, Quality Electric, is located just outside Redbone Falls on the Severn River. The dam, constructed for the purposes of the project, replaced electrical generators that had consumed almost 600 gallons of diesel fuel per day. The project, spearheaded by the Canadian Power Authority, aimed to decrease the environmental damage caused by the use of diesel fuel in the area after significant population reductions in fish, water fowl, and indigenous plant life were recorded by biologists. The plant uses water supplied from the Severn River and two tributaries to conduct, convert, and store electrical energy for home and commercial use in the general area. Because of the beautiful natural habitat in which the dam and plant were constructed, designers sought to impact the wetlands as minimally as possible. Thus, the power station and transformer cavity were built underground. Only the switching station and operations building are above ground.

The plant consists of several components.

- The upper and lower reservoirs (1.5 million cm and 1.8 cm, respectively)
- An underground power station (the function of the power station is to maintain water flow control and manage power levels) with three turbine generators

- An underground transformer cavity (which converts energy to electricity)
- A switching station
- An operations building

The plant is effectively run by five on-site people per shift. In this day of automation many plants run completely by computer.

While the safety issues important to this case could have relevance (and perhaps even larger implications) to all components in the hydroelectric power plant at Redbone Falls, the problems arise from the operations building and its staff of five people. Because the plant was developed and underwritten by several large-scale grants, the first few years of its operation require a great deal of monitoring and data collection. The staff at the plant are centrally located in the operations building, where they monitor control panels, maintain output and input levels, record fluctuations, and store all data. Because the other components to the plant are fully automated, the plant could run effectively with only one or two individuals in the operations building at a time, monitoring equipment and recording or reporting data. Because of the funding requirements, however, the plant averages five employees per shift. The automated nature of the work at the plant ensures that some employees experience a significant amount of "down time," or time spent doing little beyond waiting for something to happen. In this respect, the employees at the plant are not unlike firefighters awaiting the call that will require their specialized skills.

The operations building has three large and two small rooms. The first room houses the systems monitoring equipment, the second room houses the input/output data recording and communications equipment, and the third room serves as an information center for visitors to the plant. Just off the information center is a small break room and lavatory for employees. All equipment in the operations building is electrical.

## The Situation and Your Role

Any time an injury occurs on the job, a potential problem exists for all employees.

Six weeks ago an employee, Gwen LeChance, was seriously injured from an electrical shock that occurred when she came into contact with some exposed wiring beneath one of the control panels in the systems monitoring room. Her heart stopped, but her fellow employee, David Swain, administered CPR and she was rushed to the hospital 30 miles away. While LeChance is expected to recover, the Canadian Power Authority, upon notification of the accident, sent out a team of quality control experts to review safety procedures and instructions with employees at the plant. During the inspection the quality control team uncovered a number of major safety violations in the operations building at the plant—their subsequent survey of the power station, transformer, and switching station offered no evidence of similar violations. These violations were "below average" or "unacceptable" ratings on several points necessary to meet the electrical safety code of the Power Corporation Act of Ontario.

Clearly, the Redbone Falls plant faces a number of safety shortcomings.

The following list outlines the points of noncompliance on the initial quality control team visit.

- Two in-use electrical extension cords do not employ a grounding connector.
- Exposed wiring is present beneath one control panel.
- One metal measuring tape is located within an unsafe distance from circuit conductor.
- Disconnect switches are unmarked.
- All employees are not currently certified in cardiopulmonary resuscitation training measures.
- Three electrical enclosures are missing plates, or their covers are not fitted to regulation.
- Insufficient safety signs or crisis response processes in place for employees.

The irony of so many violations occurring at an electrical plant is not lost on the plant employees or the quality control team from the CPA. Violations of this kind have shut down larger plants. The quality control team orders plant employees to fix the problems and create appropriate instructions and signage, and agrees to send another inspection team at the end of 30 days to report on the improvements.

Your role is defined here.

You are a member of the returning quality control team sent to reevaluate the Redbone Falls Hydroelectric Plant. Since the original site visit, employees at the plant have fixed the exposed wiring and enclosure plate problems. They have also marked the disconnect switches appropriately and have grounded all extension cords. In addition, the two employees who were not certified in cardiopulmonary resuscitation are both finishing the certification course.

Despite these improvements, the employees at the plant have struggled with developing the appropriate emergency response instructions. Although they have created some basic warning signs appropriate to the area, you find that the instructions for emergency response are incomplete and that those they have finished are insufficient. When you ask the quality control team why they have had problems developing the instructions, plant employees express frustration with the task because they have little experience with developing such materials and they also see little need for them. David Swain, the spokesperson for the group, rightly points out that with so few employees on the site verbal communication is more effective in crisis situations. Too many written or even pictorial messages in the work area could ultimately serve to slow response processes if people stop to read directions.

There is potential for communication problems between plant employees and quality control team members. Why?

While you understand plant employees' arguments, you point out that creating these instructions is not optional. Regulations require the plant have instructions for response to serious injury, emergency plant shutdown procedures, and power outage response at the very least. The plant employees have made an attempt at developing two of the three procedures, but it is clear from your meetings and from the actual documents they have created that plant employees are angry at the invasion of the quality control team and have a disdain for the "hoop jumping" they face in revising the instructions.

Although the plant is technically required to comply, plant employees know there will be no recourse if they choose to comply only minimally. The hydroelectric plant project was nearly $7 million and was supported not only by the Canadian government, but by numerous foundations. While all of those sources certainly would require compliance with regulations, the plant is nearly new and it certainly won't be shut down. What plant employees seem to also believe is that they are in no danger professionally if they comply only minimally.

Your purpose is defined here.

As a quality control team member, you must review the instructions the plant employees have designed and respond to them. Your team knows that the instructions must be more detailed, but you also seek to encourage a *willing* compliance with the regulations. You would like to have the plant employees understand and agree with you about the need to create adequate instructions. Obviously, you can *make* them comply, but you know that to threaten them will not help future relations.

## Background Development

Consider the following prompts as a way to evaluate what you've read in this case. Answer the questions on paper or with others on line, or work in a group and discuss possible answers orally. You may use your answers to help you in subsequent tasks.

- The electrical plant management in this case has clearly made mistakes in terms of managing and anticipating problems. Even though it isn't old, already someone has been injured because of an accident that should never have happened. Is it possible to go too far in terms of safety preparations? Discuss your answer(s) in a small group or with your class.
- How would you characterize the relationship between the quality control team and the plant employees? Is it comparable to anything you have ever observed in your own experience?
- The quality control team is encouraging something called "buy-in" at the end. Buy-in basically means that those initiating change recognize that it is easier and more productive to work *together* than against one another. By recognizing others' objections and responding appropriately, all sides are more likely to feel they are productive components of a team. How can "buy-in" work here? Can you think of other situations where "buy-in" might have been useful to the outcome?

## Major Task 1

In the case appendix you will find two of the required three sets of instructions created by plant employees. The first is for emergency shutdown of the plant and the second is for power outage response. Plant employees have not even created the third required set of instructions—emergency response to serious injury. In a small group of three, or individually, review what the plant employees have created. Write a response to plant employees that indicates the instructions are insufficient. Remember that your aim is to encourage plant employees to see the importance of compliance while also explaining in some detail why the current instructions are flawed or limited.

## Follow-up Task

The plant employees have responded to your initial report with a request that you provide an example of a set of instructions outlining appropriate response to serious injury in the workplace. Do some research in your library, in your local industrial plants, and on the Internet to locate examples of what the plant employees have requested. Assume you will take these examples with you to a meeting with the employees to discuss what these instructions do well and how the verbal and visual design choices may or may not be appropriate to the Redbone Falls plant. Prepare a short oral presentation on why these instructions are useful models for the purposes facing the employees at the hydroelectric plant.

## Major Task 2

Plant employees have asked you to help them create their instructions for response to serious injury. In a small group, design a set of instructions for response to serious injury in an electrical power plant. You may assume that employees could risk electrocution, burns, broken bones, or choking in the workplace. Other accidents are certainly possible, but these injuries are most likely given the workplace.

## Follow-up Task

Testing is an important element in determining the effectiveness of technical instructions. Discuss with your classmates how you could test the instructions you have developed.

## Case Appendix: Emergency Procedures Instructions

The following sets of instructions are written in compliance with the regulations for industrial health and safety.

*Emergency Plant Shut-down Procedures*

In case of emergency overload, fire, or other structural damage, contact plant authorities immediately at the home base operations in Winnipeg (Speed Dial #1).

Secure the health and well-being of all employees first.

Engage control measures on main switchboard panel (lower left red button).

Await instructions from Winnipeg.

*Power Outage Response Procedures*

If outage is due to overload, check breakers located in systems monitoring room 1 (first cabinet near door).

If outage is due to weather or structural damage, contact Winnipeg operations management (Speed Dial 1).

First, disconnect all appliances and unnecessary equipment.

Second, engage back-up generators.

Await instructions from Winnipeg.

# Summary of Evaluative Criteria for Major Tasks in Case 10

| | ***1 Unacceptable***<br>Insufficient answer to assignment expectations | ***2 Below Average***<br>Inappropriate or ineffective verbal/ visual choices limit document success | ***3 Meets Task Expectations***<br>Has answered objectives of assignment, but individual components could be strengthened | ***4 Above Average***<br>Few flaws, document meets expectations, but could benefit from more attention to detail | ***5 Excellent/ Professional***<br>Few or no flaws, demonstrates keen insight into case subtleties and details |
|---|---|---|---|---|---|
| **Purpose/Key Points**<br>▪ Identifies and defines purpose<br>▪ Demonstrates audience awareness<br>▪ Summarizes details with attention to clarity and brevity | | | | | |
| **Context**<br>▪ Identifies/defines context and situational constraints<br>▪ Demonstrates awareness of document situatedness | | | | | |
| **Audience**<br>▪ Identifies/defines audience<br>▪ Establishes appropriate tone<br>▪ Understands technical details enough to communicate effectively | | | | | |
| **Organization**<br>▪ Demonstrates analytical insights<br>▪ Employs identifiable, appropriate pattern of organization | | | | | |
| **Design**<br>▪ Demonstrates awareness of visual design issues | | | | | |

CASE 11

# *Defining Ambiguous Parameters: Differentiating Technical Tools*

This case asks you to help prepare the distribution center of a small engineering firm for a review. This task requires cataloging and creating a coding system for a variety of tools and machines. You are asked to research and provide a brief technical definition for several items, including a reed-switch thermostat.

## Background

Engineers are often asked to problem-solve. Some create whole businesses or consulting jobs based on analyzing and solving others' problems.

Fox Tooling & Engineering is a small mechanical engineering firm in Pen Yan, New York. Established in 1994 by co-owners David and Peggy Fox, the firm custom designs mechanicals tools and systems for a wide variety of building needs, both commercial and private. The co-owners, both engineers, found a need to establish a business that adapted to the specific and changing needs of mechanics and builders. Too often, specific tools, even those that claimed to meet the latest demands in design and durability, could not conform to more specialized projects.

Thus, David and Peggy determined that if customer service was the key to success, they needed to meet each customer's needs face to face and one step at a time. They began to create the machines and tools people required to complete specific tasks. Although the company was small, its first few major successes with clients garnered some public attention, and within three years the business was in the black and doing well in New York and Maine. Eventually, David and Peggy determined to expand the business and hired new engineers to their staff to manage the increasing numbers of commercial and private clients.

Visualize the layout of the two buildings owned by Fox: with a partner, draw a map that breaks down the sections of the two buildings.

Physically, the firm is divided into two buildings. In the customer service area, David and Peggy meet with clients to discuss specific building needs and plans for tools that can help meet them. Once designs are finalized by the head engineers, plans and work orders go to the back of the customer service building in the "shop" area. In the shop, engineers and machinists work together to create the designs. The shop area is quite large and also includes a wide open garagelike space for testing.

The second building is the parts and distribution center. In the largest area of this building is a warehouse for parts and tools. Here engineers and machinists locate tools they need to use in order to fill work orders. If they require more specialized equipment than what they have, the chief engineer on the project places an order for it outside Fox Tooling and Engineering. In addition to the warehouse, this building includes the company's receiving and distribution center. Often large orders of parts arrive weekly. They are received at the back of this building and routed to the right team for use. In addition, most orders for clients are delivered. Therefore, when a project is completed, it is sent to the distribution building to await delivery.

## The Situation and Your Role

Your role is defined here. What kind of training do you believe you would need to be successful in this role?

You are an engineer for Fox and have worked on numerous projects with David and Peggy and on teams with other employees. You have worked for Fox for about 18 months and like the owners a great deal. You have noticed, however, that David and Peggy are engineers at heart and have had a tough time working out the bugs in their somewhat new roles as business owners.

A problem with David and Peggy's management style is decribed here.

Because they began this business shortly after they finished their degrees, David and Peggy have made a few management mistakes along the way. Although the business has prospered enough for them to hire and maintain a solid client load, it has been easy for Fox Tooling and Engineering to fall behind in a couple of areas. One thing David and Peggy have been sometimes reluctant to do, especially early on as the business was starting to blossom, is delegate authority. They both are extremely hard-working individuals, and because they love the business so much they have often chosen to overload themselves with responsibility, as opposed to relying on others to handle responsibilities with bookkeeping, employee record keeping, and warehouse inventory. Once, in an especially busy period when they were just barely keeping up with client orders, David forgot to do payroll. Fox employees went three extra days without a paycheck, and then when they did receive them, they quickly discovered that David forgot to deduct the monthly insurance costs.

It is not unusual, then, that David and Peggy never hired a warehouse manager or inventory clerk. The responsibility intially was Peggy's; when business became brisk, however, Peggy assumed more and more responsibility in client relations and project development—both of which were really her expertise to begin with. When shipments of inventory would arrive, often whoever answered the back door would sign for the shipment and leave the inventory stacked along the wall of the warehouse, presumably until Peggy could get to cataloguing it and putting it in the right place. An engineer desperate for a part or tool would typically storm back to the warehouse, angrily stuff materials from boxes in the general vicinity of where they belonged, and return to work, frustrated at the waste of time.

Employees had spoken with Peggy and David, together and separately, about the problems with inventory management, but nothing changed for very long.

After a complaint, Peggy or David would scramble to try and put order to chaos in the warehouse, but in a few weeks the priorities of the design and project management would take over and the inventory would fall behind again. That is, until Peggy thought she'd lost a very important, very expensive magnet.

Your friend and fellow employee Franklin Shabazz puts his hand on your shoulder one afternoon and says, "Have you heard we've got an all-employee meeting at 4?"

You shake your head no. "What's it about?"

Franklin shrugs. "All I know is that Peggy is frantic over some magnet she thinks is lost. I think she ordered it for that NYU job and it's something like $4,000. Anyway, delivery service says they delivered it, and it's nowhere to be found. That's one expensive loss."

You agree and ask, "What kind of magnet is it?"

"I don't know," Franklin admits. "I think Peggy has kept that pretty close to the chest. It's her project, so naturally she's probably the only one who knows what it's for. Anyway, I'd bet this meeting is just about doing some sort of massive search."

Based on the exchanges in this case, how would you say the coworkers get along with each other? With David and Peggy?

Later that afternoon at the meeting everyone is gathered in the break room. Most have speculated that the meeting is about the missing magnet. David stands and clears his throat. "I imagine all of you have heard by now about Peggy's missing magnet." Though it's a serious loss, there are a few chuckles among those at the meeting who acknowledge how absurd the loss sounds. David continues, "Well, we found the magnet a little while ago, just where I figure most of you assumed it would be—in the warehouse."

Karna Evers, sitting at the next table, interrupts. "Oh! Was it behind that stack of about twelve boxes along the east wall?" The room erupts in laughter. The warehouse is simultaneously a running joke as well as a constant nuisance to the employees at Fox.

David good-naturedly laughs along with the rest of the group, then adds, "Laugh while you can, folks. We're all going to be doing something about that warehouse over the next few days." At that point David explains that he and Peggy have finally come to the conclusion that the warehouse must be managed by someone other than themselves, and in fact, after the magnet fiasco, they have both realized that unless they do something soon, Fox may potentially lose business.

"So," Peggy concludes, "we've decided we're going to hire a warehouse manager who can also serve as the chief inventory clerk for the business." The staff applauds.

"But there's a catch," David adds. "When we were in there this afternoon, it became clear pretty quickly that unless we straighten things out in there, we won't be able to pay someone enough money to take that job. It's no news to the rest of you, I know, but that place is a disaster.

"Therefore, we're closing up shop tomorrow, Friday. We'd like to spend the next three days working on putting the warehouse in order, taking inventory of what we have, and come up with a list of what we have, what it's used for, and where it goes. Once we get things at least semi-organized, we can hire someone to take it over permanently. We can't require you all to be here this weekend, but we'll pay you time and a half. And I think the rewards will extend beyond money, once we're finished."

Your purpose is defined here. What are your challenges?

You are among a dozen employees who agrees to help with the inventory project in the warehouse alongside Peggy and David. Each person is assigned a specific space in the warehouse and asked to take a clipboard and pen and name and count every item located in the area, while restacking and generally putting things in logical compartments. Once you have named each item and provided a count, you are to provide a brief description (no more than 10–15 words) of the item's purpose. If you have found items in your section that do not belong there, you are to label them, write a technical description of their use, and attach it to the item(s) which will be placed in the appropriate location later.

## Background Development

Consider the following prompts as a way to evaluate what you've read in this case. Answer the questions on paper, with others on line, or work in a group and discuss possible answers orally. You may use your answers to help you in subsequent tasks.

- Consider what it means to "custom design" tools and systems to fit specific commerical and private building needs. What sorts of skills might an engineer need to accomplish this task? Discuss with classmates.
- When professionals refuse to delegate and spread work out among team members, a business or organization can suffer. Have you ever been involved with a group or been employed by a company that faced this problem? As a class, discuss the specific problems such organizations face when only a few people shoulder the work responsibilities.
- Taking inventory in the Fox warehouse sounds like quite a job. With a partner, create a list of potential problems employees may face with this project, based on your knowledge of the business and its owners. If you could recommend anything to make the process run smoothly, what would it be and why?

## Major Task 1

In the case appendix you will find a basic list of inventory items and their counts from the section you have been assigned to cover in the warehouse. The section you are dealing with, though important, is not normally filled with overly technical items. You are working with a section of the Safety Products and Tools area. The items listed are not difficult to describe, but you must be precise and accurate in your description. If there is an item on the list that you do not know or have a description for, research the item in your library or on the Internet and locate a viable description for it given the context descibed in the case.

Using the list as a starting point, create a catalog that offers a logical organizational pattern for a reader, outlining the names of the items, the counts, and their descriptions. You may assume you have three shelves 4 feet long each and

6 feet tall. Each set of shelves has three levels. You also have three 12' by 3' tables at your disposal upon which to organize the items. Have your catalog reflect the organization you envision for the items. A reader should be able to look at your inventory and know where to go in that section to locate a specific item.

## Follow-up Task

In your section, you have uncovered several items that do not belong among the Safety Products and Tools. One of those items is a reed switch thermometer. Conduct some research into the purpose of a reed switch thermometer and create a technical description of its purpose in 50 words or less. In your technical description, you should focus on explaining how it works and what it is used for. Then, using the list of sections in the warehouse, determine where the reed-switch thermostat ought to be placed. Include your recommendation in bold type at the conclusion of the description.

## Major Task 2

Peggy and David have hired Justin Ortiz to manage the warehouse and inventory. In a memo to Ortiz, explain the logic behind your organizational strategy and offer to provide any further insights upon his request.

## Case Appendix: Cataloging Technical Tools

**List of Warehouse Categories**

- Bandsaw blades
- Coolants/Lubricants/Adhesives/Abrasives
- Cutting tools
- Machine tools
- Machine tool accessories
- Precision tools
- Safety products and tools
- Industrial supplies
- Tables and bases
- Coil cradles and reels
- Computer equipment and keyboards
- Electronics and wiring
- Press shop equipment
- Miscellaneous/Special order

**Safety Products and Tools Inventory List**

| | |
|---|---|
| Ear plugs (plastic) | 9 prs. |
| Face shields | |
| • 1/2 face (metal) | 4 |
| • Full face (metal) | 6 |
| • 1/2 face (plastic) | 2 |
| • Full face (plastic) | 1 |
| First aid kits (basic) | 3 complete, 1 partial |
| Hard hats | |
| • Hard plastic | 5 |
| • Metal | 8 |
| Knee pads | |
| w/ velcro tie | 10 |
| w/ leather straps | 1 |
| Elbow pads | 3 |
| Face masks (for oxygen) | 8 |
| Safety glasses | |
| w/ strap | 10 |
| wrap around | 20 |
| Safety mats | |
| • rubber | 4 |
| • turf | 6 |
| Welding curtains | 4 |
| Welding goggles | 8 |

| | |
|---|---|
| Gloves | |
| • rubber | 8 |
| • insulated | 4 |
| • net | 2 |
| Wrist stabilizers | 1 |
| Carbon monoxide detectors | 2 |
| Radon test equipment | 2 boxes |
| Fire safety equipment | 3 boxes |

# Summary of Evaluative Criteria for Major Tasks in Case 11

| | *1 Unacceptable*<br>Insufficient answer to assignment expectations | *2 Below Average*<br>Inappropriate or ineffective verbal/ visual choices limit document success | *3 Meets Task Expectations*<br>Has answered objectives of assignment, but individual components could be strengthened | *4 Above Average*<br>Few flaws, document meets expectations, but could benefit from more attention to detail | *5 Excellent/ Professional*<br>Few or no flaws, demonstrates keen insight into case subtleties and details |
|---|---|---|---|---|---|
| **Purpose/Key Points**<br>▪ Identifies and defines purpose<br>▪ Demonstrates audience awareness<br>▪ Summarizes details with attention to clarity and brevity | | | | | |
| **Context**<br>▪ Identifies/defines context and situational constraints<br>▪ Demonstrates awareness of document situatedness | | | | | |
| **Audience**<br>▪ Identifies/defines audience<br>▪ Establishes appropriate tone<br>▪ Understands technical details enough to communicate effectively | | | | | |
| **Organization**<br>▪ Demonstrates analytical insights<br>▪ Employs identifiable, appropriate pattern of organization | | | | | |
| **Design**<br>▪ Demonstrates awareness of visual design issues | | | | | |

# CASE 12 *Reviewing Royal Built's Owner's Manual: The Antilock Braking System Section*

The task here is to review the description of an antilock braking system in an owner's manual for a new automobile. The description must be accessible to the lay reader, but also must be complete and accurate. You are asked to read, revise, and report on the changes made to the existing manual description.

## Background

This paragraph should give you insight into Royal Built's public image. If you were to assign specific adjectives to Royal Built, what would they be?

Royal Built is a U.S. automobile manufacturer based in Dallas, Texas. Established in 1952 by Henry Patten, Royal Built has enjoyed a long history of high-quality American auto production. Unlike the megacorporations Ford, General Motors, and Toyota, Royal Built has concentrated solely on developing a relatively limited variety of cars and trucks. Limiting production to just a few models has allowed Royal Built to perfect designs. The models include one compact model, two sports sedans, one four-door sedan, and one luxury four-door. In fact, even though the cars are not as expensive as their luxury counterparts, Royal Built autos are consistently rated among the best built cars in the United States, and several models are compared with Germany's Mercedes-Benz for performance and luxury.

The DOS system is brand-new technology designed specifically for consumer safety.

Royal Built comes out with a new model every four to six years. The company's latest model, introduced in 1995, was called Applause! Consumers and critics agreed that Applause! was the finest vehicle Royal Built ever engineered and its sales have proven this fact. One of the most interesting design features of Applause! is the way the antilock braking system (ABS) works in conjunction with the Direct Object Sensor (DOS), the last word in safety technology. Royal Built's Applause! is the first auto to incorporate and market the device. The DOS system works with the ABS system to detect objects directly in the path of the vehicle, warn the driver with a flashing light on the dash, and brake automatically before impact if the driver does not respond to the warning.

Royal Built's Applause! marketing campaign strategists incorporated many emotional testimonials from Applause! owners who had been "saved" by the DOS/ABS system. One particularly effective commercial featured 12-year-old

Mara Suzanne, who told the story of her father's return from an extended business trip to Asia "just so he could get home in time for my birthday party." But because he was so tired from jet lag, she explains to audiences, he started to fall asleep while he was driving. As the car careened off the road, the DOS/ABS system responded to the tree the car was rushing to meet. Mara Suzanne's father was saved by the car's ability to stop a foot from the tree and what could have been a fatal impact. The commercial concludes with Mara and her father smiling together as Mara says, "He didn't make my birthday party on time, but he sure gave me the best present—himself!"

Why is it useful to acknowledge limitations to new technology?

The DOS/ABS system obviously cannot prevent all accidents, however. In the car's description, Royal Built and the DOS patent holder, Fail-Safe, explain that the DOS/ABS system can warn and respond to objects that are stable (unmoving, like a tree or a guardrail or an obstruction in the road), or those that are moving away from the Applause! vehicle but at a slower rate (a vehicle directly in the path moving slowly enough to risk a rear-end collision). However, the DOS/ABS system cannot prevent head on collisions (when the oncoming vehicle is also moving forward to meet the Applause! driver's vehicle head on), nor can it sense obstructions that range from beyond the edges of the grille of the Applause! auto (a vehicle coming from the driver's side to broadside).

Historically, Royal Built's owner's manuals have tried to educate drivers about the auto's advanced systems and part names. Critics have claimed that the technical language was too difficult for the average consumer. Royal Built has responded that they refuse to talk down to their consumers; however, they include instructions for use that are less technical alongside the descriptions.

## The Situation and Your Role

Your role and purpose are defined here.

You are a technical writer trying to land a job at Royal Built. Royal Built's employment record is admirable, and you have confidence in the product.

As part of the screening process for hiring, Royal Built asks prospective employees to rewrite a component to the owner's manual for the Applause! automobile. You are to use the existing technical description of the system as well as the instructions to help you.

Two supervisors disagree about the purpose of technical communication. How does the conflict affect your approach?

The woman interviewing you is Grace Paul. Ms. Paul is head of the division of technical communication and public relations. Though technical communication and public relations are under the same umbrella heading at Royal Built, Paul's background and strongest affiliation is in public relations. Paul is a long-time advocate of simplifying the public messages (especially in owner's manuals and repair manuals) and has made this abundantly clear throughout your interview. It has also become clear that Paul does not see eye to eye with Carl McFadden, the director of technical communication services at Royal Built. You have also interviewed with him, and he made it clear that "the public is not stupid." According to McFadden, Royal Built wants its consumers to invest not just in the product, but in the knowledge and integrity behind the product. In his view, technical communication shouldn't be watered down.

You understand that there is a basic difference in approach between these two individuals. Although Paul is technically McFadden's superior, the differences seem not to threaten McFadden's employment status with Royal Built. You suspect that there are a number of political issues at work here, but because you are an outsider, you can ascertain nothing concrete.

You face space and time constraints.

Your writing sample, which you write between interviews in a little computer room off Paul's office at Royal Built, will be read by both McFadden and Paul. While McFadden's input is influential, the hiring decision is ultimately Paul's responsibility.

## Background Development

Consider the following prompts as a way to evaluate what you've read in this case. Answer the questions on paper, with others on line, or work in a group and discuss possible answers orally. You may use your answers to help you in subsequent tasks.

- When new technology is incorporated in a commercial product, it is extremely important to consider how to communicate its function and limits. Have you ever observed a new product's introduction to the public? How did the manufacturers caution consumers about limitations? Discuss your answers in a small group with classmates.
- How would you define your purpose in this case? You may have a number of purposes to deal with. Can you rank them in order of importance?
- Although technical communication services and public relations fall under the same heading in Royal Built's case, the two divisions seem to be at odds about the nature of the audience. Do you think this might be a common difference of perspective between public relations communication and technical communication? Discuss as a class how the aims of such communicators are similar and different, based on your knowledge of employers who require the services of both.

## Major Task 1

Using the case appendix materials (a technical description and a set of instructions for the antilock braking system), construct a one-page description and list of tips for the owner's manual for Royal Built's Applause! automobile. Grace Paul has explained that you need not worry about following any specific convention for design—she is more interested in seeing the choices you make on your own initiative. She has also explained to you that the materials you use for your test are not exceptional; they are average descriptions, though not the ones Royal Built uses currently. This is simply a test to see how you communicate the information.

You are allotted one hour to compose the draft of this text. Your computer, located in a room just across the hall from Grace Paul's office, is fully equipped with word processing programs and Internet access. You may choose to use some of your composition time for research if you wish.

## Follow-up Task

Grace Paul has reviewed your writing sample and now, along with Carl McFadden, has asked you to give a 10-minute oral presentation explaining your rhetorical choices offering insight into how you would treat the assignment differently if you were allotted a few days to complete it. Remember that this is still part of your interview process and you are still trying to land this position. Your response to the task should be persuasive as well as analytical and informative.

## Major Task 2

Grace Paul is intrigued with your responses in Major Task 1 and the Follow-up Task. She has one more test for you. She would like to see how you might change the text you created for the Royal Built website. She doesn't want you to look at the website to get ideas; rather, she's interested in how creative you can be about communicating technical information on the Internet. For this task you may choose to present your ideas orally (10–15 minutes) or actually create your document electronically.

# Case Appendix: Description and Instructions for an Antilock Braking System

## Technical Description

An integral anitlock braking system (ABS) is powered primarily by a singular unit that consists of the brake master cylinder, high-pressure pump, and accumulator (or modulator). The power unit actively engages the ABS with the power-assist braking system and the Direct Object Sensor (DOS). The components in the system include a master cylinder, a modular valve assembly that controls the brake line pressure to each wheel, speed sensors that measure wheel speed, and two electronic control units—one on board and one outer board. The outer-board electronic control unit senses and reports potential obstructions outside the vehicle. The on-board electronic records vehicle speed and during brake locking signals the hydraulic actuator, which applies and releases the brakes as many as 15 times per second. If the electronic control units determine that a wheel is decelerating too rapidly compared with the other wheels, it activates the the hydraulic actuator to keep the wheel rolling. The two electronic control units also interface to respond to potential direct object interference.

## Owner's Manual Instructions

The antilock braking system (ABS) is an automatic characteristic of your vehicle and cannot be disabled manually. It is designed to maintain stopping and steering control and actively prevents rear wheels from locking and skidding.

The ABS system engages depending on the amount of traction your vehicle's wheels experience. Traction is often determined by pavement and weather conditions, but may also be affected by speed. The sensation you will experience with ABS engagement is a slight pulsation in the brake pedal. This pulsation may occur automatically when you actively brake.

In all conditions, you should gauge the force with which you engage the brake pedal with the response from your vehicle. The ABS system does not guarantee safe driving in all road conditions, and Royal Built strongly encourages you to use good judgment in determining safe driving conditions.

You may feel slight pulsation in the brake when you start your vehicle. This is an indication that the the ABS system is engaged and working. If the red light marked ABS on your instrument panel flashes, this is an indicator that your ABS system should be serviced by your dealer. The brakes in your vehicle should still function properly and provide normal stopping ability, but the ABS function may be compromised.

Stopping on loose or uneven surfaces may require longer stopping distance allotment with an ABS vehicle.

# Summary of Evaluative Criteria for Major Tasks in Case 12

| | *1 Unacceptable*<br>Insufficient answer to assignment expectations | *2 Below Average*<br>Inappropriate or ineffective verbal/ visual choices limit document success | *3 Meets Task Expectations*<br>Has answered objectives of assignment, but individual components could be strengthened | *4 Above Average*<br>Few flaws, document meets expectations, but could benefit from more attention to detail | *5 Excellent/ Professional*<br>Few or no flaws, demonstrates keen insight into case subtleties and details |
|---|---|---|---|---|---|
| **Purpose/Key Points**<br>▪ Identifies and defines purpose<br>▪ Demonstrates audience awareness<br>▪ Summarizes details with attention to clarity and brevity | | | | | |
| **Context**<br>▪ Identifies/defines context and constraints<br>▪ Demonstrates awareness of document situtedness | | | | | |
| **Audience**<br>▪ Identifies/defines audience<br>▪ Establishes appropriate tone<br>▪ Understands technical details enough to communicate effectively<br>▪ Shows initiative and creativity with electronic design features | | | | | |
| **Design**<br>▪ Demonstrates awareness of visual design issues | | | | | |

CASE 13

# *Evolution Publishing: Finding a New Way to Communicate Technical Information*

Evolution Publishing is a small educational publishing house that specializes in scientific texts and magazines. Looking to expand its market share, top management seeks to begin a lower-level science magazine and asks you to brainstorm for different ways to communicate a technical process. One task asks you to create a cartoon strip to explain a process to young readers.

## Background

Recall the science-oriented magazines you have seen previously. Consider what their intended audiences were.

Evolution Publishing is an academic publishing firm located in Seattle, Washington. Established in 1974 by biologist and former University of Alaska professor Wilcox Esterhaus, Evolution began by publishing a limited number of specialized biology, chemistry, and physics texts for advanced graduate and researcher audiences. In the early 1980s, Evolution hired two technical communication specialists with training in biological sciences. As a result, the publishing firm also began producing secondary school biology textbooks and established a national reputation with the bestselling *Biology Made Easy.*

Largely because of that text, Evolution continued to grow, and in the late 1980s the company began publishing two science-oriented popular magazines called *Science and You* and *The Physical World.* While both magazines target a general popular magazine audience, they have done particularly well in classroom sales. Recent marketing strategies have targeted secondary and postsecondary science instructors with great success. Sales consultants have encouraged teachers to require the magazines as supplemental texts for their students, and Evolution has followed through with teaching tools packets for those instructors who adopt the magazines. The appeals have been successful, and sales figures for these magazines have experienced a significant leap in the last three years. Sales projections anticipate continued growth, which is, of course, good news for Evolution Publishing.

Wilcox Esterhaus died last year. The new editor-in-chief and operations director is Esterhaus's daughter, Janelle Esterhaus-Cronn. Sam Kennedy, formerly sales director, moved into Esterhaus-Cronn's previous position of development editor. While both have worked in the business for at least 15 years, they have different visions for what sort of image they want to communicate for Evolution.

The conflict between Janelle and Sam is defined here. What is the source of their differences?

Janelle has a strong desire to place more emphasis on textbook development and advanced scientific research. Most employees suspect that at least a part of this perspective is undergirded by a desire to pay tribute to her father's vision for the publishing firm because he was a scientist and researcher himself. Sam comes from a marketing background and strongly believes that the future for Evolution lies in developing appreciation for the sciences at a very young age. The popular press magazines were Sam's concepts, and Wilcox Esterhaus was very supportive of the endeavors, in part because they brought in good money and helped advance Evolution's reputation.

## The Situation and Your Role

Sam's purpose for calling the meeting is defined here.

Sam has called together a team of four editors and writers to help craft a proposal to Janelle Esterhaus-Cronn and board members for the development of a new popular magazine produced and published by Evolution staff. His idea is to move into the popular press with a science magazine that targets young readers ages 4–12.

Your role in the case is defined here. What do you see yourself doing in this role?

You are a technical communication specialist with some background in marketing, and Sam has asked you to be a part of the development team on this project along with Chris Charters, the technical editor for *Biology Made Easy;* Melissa Story, a technical consultant and writer of a monthly syndicated column on women's health issues; and Tyrone Brown, a graphic artist and design specialist.

"This is probably going to be an uphill battle, folks, " Sam says at your opening meeting. "It's pretty clear that Janelle is not terribly interested in maintaining this focus on magazine/popular press development. But I think this is a logical move. A science magazine for youth could go like hot cakes if the preliminary sales forecast is on target."

The team's overall goal is defined here. What are the obstacles to reaching that goal?

Sam has asked your team to develop a list of 10–15 ideas for different sections that might be included in a science magazine for youth and to write them up with explanations in a brief report to him. The two ideas Sam seems already committed to, however, include a comic strip that deals in a humorous manner with some scientific process and a regular short story that addresses "mysteries of science." You should assume that these ideas will be included in your list for Sam. Together, then, your team will determine which of the ideas seems most feasible and creative and from a shortened list will develop a mock-up of the magazine as a part of the proposal process addressed to Janelle Esterhaus-Cronn and the rest of the board.

## Background Development

Consider the following questions as a way to evaluate what you've read in this case. Answer the questions on paper or with others on line, or work in a group and discuss possible answers orally. You may use your answers to help you in subsequent tasks.

- How would you describe the differences between Sam and Janelle? Make a side-by-side list of the characteristics of both people. Based on what you know, how might Sam successfully appeal to Janelle to win her approval?
- Compare three or four of your favorite cartoon strips. Make a list of the similarities they all share. Consider their purposes, their appeals to humor, the kinds of audiences they attract. What qualities do they share? How are they different? Is there any educational purpose you can detect in these cartoons?
- In a small group, discuss what strategies you might use in appealing to a skeptical or outright hostile audience.

## Gathering More Information

With a partner, scan a variety of popular magazines targeted at both adults and children (some examples might include *Discover, Earth Magazine, Scientific American, Popular Mechanics, Cricket, Ranger Rick, Crayola Kids Magazine,* and *Highlights*). Look at the different types of articles, stories, graphics, and advertisements they offer, and from this try to develop at least 10 possible ideas for presenting technical and scientific information to young readers. Another strategy for research might include talking with elementary school teachers or early childhood specialists who have insights about how best to communicate technical information to children. With your partner, summarize your ideas and clip examples or cite sources so you can support those ideas later.

## Communicating Your Research

Based on your research, write a memo to Sam detailing your ideas and offering brief examples of how technical information might be articulated in them. Highlight those ideas you believe are the best, since you will ultimately need to have a focused appeal when you take your ideas to Esterhaus-Cronn.

## Major Task 1

Sam loved the team's ideas and has asked you to help develop them for a formal presentation. Again with a partner, develop a cartoon strip that illustrates a technical process or a mystery of science. You may assume that you'll present your ideas orally in a meeting with other team members and Sam, but not at this point to Janelle or the board members. Your facts need to be accurate, but you should feel free to be creative and choose whatever process appeals to you. If you don't see yourself as an artist, that's fine. Think about your options for creating visuals. There are many computer programs today that offer help with developing visuals such as these. In addition, you may feel free to hand-draw your cartoon, knowing that when the strip is finalized, it will be done by a graphic artist. Create the visuals so that the reader has an understanding of what's taking place. Include dialogue or background explanation to make your points about the technical or scientific process clear.

## Major Task 2

Using the draft of the cartoon strip you developed for the oral proposal in Major Task 1, write an accompanying argument (no more than 250 words) for why your idea will appeal to both children and adults and how it is educational in nature. You may assume that either you or Sam will present all or part of this argument orally to Esterhaus-Cronn.

## Follow-up Task

Tyrone and Chris have offered several interesting and unique suggestions, but one of their cartoon strips deals with ovulation. You think that their sense for the age group is off target, and offering the cartoon strip about ovulation as an example might jeopardize the team's overall appeal to an already skeptical audience (Janelle and the board members). Send a brief memo that outlines your concerns to Chris and Tyrone.

# Summary of Evaluative Criteria for Major Tasks in Case 13

| | ***1 Unacceptable***<br>Insufficient answer to assignment expectations | ***2 Below Average***<br>Inappropriate or ineffective verbal/ visual choices limit document success | ***3 Meets Task Expectations***<br>Has answered objectives of assignment, but individual components could be strengthened | ***4 Above Average***<br>Few flaws, document meets expectations, but could benefit from more attention to detail | ***5 Excellent/ Professional***<br>Few or no flaws, demonstrates keen insight into case subtleties and details |
|---|---|---|---|---|---|
| **Purpose/Key Points**<br>▪ Identifies and meets purpose<br>▪ Articulates key points clearly<br>▪ Summarizes key issues with attention to clarity and creativity | | | | | |
| **Context**<br>▪ Identifies/defines context and situational constraints<br>▪ Demonstrates awareness of cartoon's situatedness | | | | | |
| **Audience**<br>▪ Identifies/defines audience and aims to meet identifiable needs<br>▪ Establishes appropriate tone<br>▪ Understands technical details enough to communicate effectively | | | | | |
| **Oral**<br>▪ Presents cartoon with enthusiasm and full understanding of its purpose<br>▪ Demonstrates attention to visual components of presentation as well as oral (has the team member used approprite equipment/technology to aid audience in accessing and understanding ideas?) | | | | | |

*continued on the following page*

## Summary of Evaluative Criteria for Major Tasks in Case 13, *cont.*

| | ***1 Unacceptable***<br>Insufficient answer to assignment expectations | ***2 Below Average***<br>Inappropriate or ineffective verbal/ visual choices limit document success | ***3 Meets Task Expectations***<br>Has answered objectives of assign-ment, but individual components could be strengthened | ***4 Above Average***<br>Few flaws, document meets expectations, but could benefit from more attention to detail | ***5 Excellent/ Professional***<br>Few or no flaws, demonstrates keen insight into case subtleties and details |
|---|---|---|---|---|---|
| **Design**<br>▪ Demonstrates awareness of visual design elements of task<br>▪ Demonstrates an awareness of design options & technological aids in the development of these options | | | | | |

# PART IV

*Completing Documents*

CASE
14

# *Meeting New Safety Standards: Dreamscape's Fire Escape Plan*

You are asked to work collaboratively to create technical documents that detail instructions for emergency measures in case of fire. You will create visuals for posters, technical instructions for walls, and a memo detailing changes to the state inspectors.

## Background

Consider the importance of understanding the history of a place when you communicate with clients or customers. How important is it for technical documents to instill "faith" in users?

In 1994, a popular coffeehouse in east Los Angeles caught fire when wiring in the kitchen area short-circuited and led to a blaze that spread quickly through the back third of the establishment. A waiter was killed trying to douse the blaze, and nine customers and employees were injured as they tried to fight their way out the front doors of the crowded gathering spot. In an investigation following the blaze, the fire marshal reported that the owners of the coffeehouse had failed to provide appropriate alternative routes out of the coffeehouse in case of fire and had also failed to post the appropriate fire escape signs and directions. While a sign near the front entrance indicated that the door could be used for escape in case of fire, the directions for second-story escape had somehow been "lost," according to the owners. In addition, the technical instructions for employees for evacuating customers and dealing with a fire were located in a file drawer in the back office, and none of the employees on duty the day of the fire was aware that the instructions even existed.

Shortly after the fire marshal's report, the family of the deceased waiter filed suit against the coffeehouse owners for negligence. Two others who were injured in the blaze filed shortly thereafter. Within three months after the blaze, the owners of the coffeehouse filed for bankruptcy and closed the business with only minimal repairs completed on the fire-damaged building.

Why is it useful to know that Whitaker and Franklin have followed the lead of city officials in revitalization efforts in the Los Angeles neighborhood? How might such circumstances affect the nature of

Early in 1996, almost 14 months after the coffeehouse fire, Malik Whitaker and Hosea Franklin purchased the building during a revitalization effort spearheaded by city council members and local neighborhood activist groups. Whitaker and Franklin were interested in turning the two-story, 27,000-square foot building into an upscale video store/coffee shop. Dreamscape was designed to appeal to professionals and family-oriented clientele (in keeping with the city council's revitalization efforts in the area), so owners refused to carry porno-

the professional communication the business initiates, both internally and externally?

graphic or excessively violent films. Instead, their extensive rental collection would feature a large collection of popular movies as well as hard-to-find older films and documentaries and new laser disc videos. One of the most interesting elements of the proposal they offered investors was an Internet library access area that allowed customers to search university archives and film warehouses all over the world for out-of-circulation or hard-to-get versions of films. Dreamscape would then offer to place orders (usually for purchase rather than rental) for customers.

Franklin and Whitaker proposed to make this business venture anything but a "standard" video rental store. In fact, Dreamscape was designed to offer two unusual features. First, in addition to the standard rentals and sales, Dreamscape would also rent three private viewing areas for small parties on its main floor. These private screening areas were soundproof rooms that would seat up to 15 people in comfortable couches and lounge chairs around a large screen television. The rooms would be useful for small university classes, children's birthday parties, clubs, and sports fans, among others, and would rent for approximately $20/hour, plus the cost of the video rental. Customers could not bring in their own privately owned videos, and no alcohol would be allowed in the private screening rooms.

The second unusual feature of the Dreamscape proposal to investors was that in addition to the commercial video store component of the business, Whitaker and Franklin envisioned an "Entertainment and Eats Nook" that would offer a gourmet snack menu, soft drinks and imported coffee, and occasional live entertainment. Whitaker and Franklin presumed that shoppers, particularly during holiday seasons, would be interested in a shop that offered a gathering and resting spot.

Prospective investors were impressed with the comprehensive proposal that Whitaker and Franklin offered and were pleased at the commitment both seemed to have to the east Los Angeles neighborhood. Dreamscape offered high-tech entertainment within an environment that would be an appropriate gathering spot for children, professionals, students, and area residents of all ages. Whitaker and Franklin also had a proven track record with successful business ventures. In 1986, they built a small computer software design company. The company, which employed approximately 32 people in Oakland, California, had purposefully stayed relatively small (adding merely 14 positions in eleven years), and cultivated personal relationships with its clients. It was regarded as highly successful for its small-scale approach and boasted some of the leading programmers and software consultants in the area. As a result of the well-articulated proposal for Dreamscape, Franklin and Whitaker's proven business acumen, and an idea that fit well within the larger revitalization scheme for the city, Franklin and Whitaker had almost no trouble guaranteeing investment money in the project.

Note that a history of inadequate signing and inaccessible technical instructions caused problems for the previous owners. In planning safety

Once plans for the new business were approved, city officials informed Whitaker and Franklin that one of their first priorities must be to address building code issues. Because the building had remained empty for nearly two years, and because the renovations following the original fire had only minimally begun when the previous owners had filed for bankruptcy, Whitaker and procedures, do you believe it is possible to have too much in terms of signs and/or directions? Discuss with classmates.

Franklin would need to work closely with building and fire inspectors to determine what changes were necessary to bring the building up to safe standards. The business partners agreed, and during their early conferences with building code inspectors and the fire marshal, Whitaker and Franklin learned of the previous owners' problems with inadequate signing and inaccessible technical instructions. At this point in the development plans Whitaker and Franklin agreed to make sure that the fire escape and earthquake emergency measures were communicated clearly at Dreamscape so that they would not be creating the same potentially dangerous conditions as the building's previous owners.

## The Situation and Your Role

The renovations for the Dreamscape building are near completion. Whitaker and Franklin have hired personnel and are in the process of planning the grand opening week for October 1998. Originally, Whitaker and Franklin tried to create fire escape and earthquake signs and instructions. The building inspector, however, informed the owners that such a job was better managed by a professional, someone who understood building codes and technical instructions.

Franklin and Whitaker consult with their writers in Manuals and Instructions at their software company, TechImage. One of their writers, Ethan Wickham, knows that you have done some freelance writing and design and puts the owners in contact with you. Although you have never designed fire escape plans before, you have some good experience with writing technical instructions.

You work for Four Brothers Printing as a production assistant, and your primary work involves brainstorming with clients to develop appropriate visual/verbal designs to meet specific rhetorical needs (e.g., a company logo, brochure, glossy cover for annual reports, etc.). You take the rough product, finalize it, obtain approval from the client in a subsequent meeting, and deliver it to the printing department at Four Brothers. While you have only worked with Four Brothers for 14 months or so, you have known Isaac Josefowicz, the eldest in the brother partnership, for more than 10 years. Isaac is an excellent print shop foreman and his work is above reproach. You work well together, and Isaac has given you some important insights into the printing business.

While you enjoy your work at Four Brothers, it is not what you originally set out to do professionally. You have a B.S. degree in industrial technologies management systems; however, because the job market was flooded when you graduated and you are geographically limited to the greater Los Angeles area for personal reasons, you have been unable to secure a permanent full-time position in the industry. Isaac is aware of your professional goals and suggested shortly after you began working for him as a Four Brothers production assistant that you consider doing some freelance consulting in the area. "OK, so you won't be managing a new systems operation right away," he pointed out wisely. "You'll probably get called to set up high school computer labs or something—depending, I suppose, how you advertise yourself. But eventually your name will get out there as someone who understands and can communicate about industry and technology. Little jobs can sometimes lead to big ones." Isaac had graciously allowed you to use Four Brothers equipment (on your own time) to get started, and after about eight months you had a few startup clients who were at least paying you enough to help you begin to put money aside for your own advanced computer system at home. Over the past eight months, you have found that you enjoy the challenge of technical communication more and more; therefore, your interests are moving in the communication management direction.

Do you feel there is a possible conflict of interest here?

Your prospective role is defined here. What do you anticipte doing in this role?

You are slightly surprised, however, by Ethan Wickham's phone call telling you he's recommended that his bosses, Whitaker and Franklin, approach you with the freelance job creating the signs and technical instructions for Dreamscape's fire escape and earthquake procedures. While you feel confident you can create what Franklin and Whitaker may need, such a project is really more along the lines of what Four Brothers would contract for. Therefore, this feels like a slight conflict of interest—the first you've run across in your freelance business.

"I think maybe this could be a conflict of interest, Ethan, though I appreciate you thinking of me for something like this," you tell him over the phone. "Could you suggest to your bosses that maybe Four Brothers might be a good bet? I'd work as the production assistant on the project. That way, they're still working with me, but I'm not stepping on any toes."

"Well, sure I could recommend they talk to your boss. What's his name? Josefowicz? But I'm afraid it might be a little late. I tossed your hat in the ring yesterday, and I suspect they'll be calling today or tomorrow. You can always tell them yourself that you feel like this might be a conflict of interest," Ethan suggested.

"Sure, I guess I could."

Ethan added, "The reason I thought of you, though, is because I knew you were interested in maybe getting into some sort of information technology, and I think TechImage might have a position opening up. It's entry level, and I don't know what all it entails, but Franklin and Whitaker are pretty good guys to work for. It might be a good step."

Indeed. TechImage might offer an interesting opportunity professionally. But as good as Isaac has been to you over the past 14 months, you certainly don't want to burn any bridges. You call Isaac shortly after your conversation with Ethan. Isaac agrees that this is as close to conflict of interest as anything you've done to this point, and Dreamscape would be a potentially big client for Four Brothers if its commercial potential is realized. "I think," said Isaac, "that it's likely Mr. Franklin and Mr. Whitaker probably want to spend the least amount of money possible on this job. Most of the time you pay out less money if you're working with a single consultant as opposed to a full business."

You agree and tell Isaac that if Franklin and Whitaker call, you will turn them down. "Wait, though," Isaac cautions. "We might still be OK, if you make it clear to them that beyond this one project they should think about Four Brothers as their primary printing service. It's better to have *some* connection than to send them packing to another professional completely unrelated to us. What I'd ask is that you offer it up through Four Brothers. If they don't want to go commercial, fine. Take the job—if you want it, that is—and make sure you indicate that this is exactly the kind of thing you do for us and you'd hope they use us in the future. At least we maintain some connection there, anyway."

You hang up with mixed feelings. On one hand, you entered into consulting with the intent to develop your expertise in industrial technologies system management. The consulting has led you instead to develop an interest in technical communication. The projects you have landed are directing you into a field that may put you in immediate competition with Four Brothers (though, admittedly, scale would certainly differentiate you at this point). On the other hand, perhaps your change in focus has simply arisen from time and distance away from your original field. If the Dreamscape job could open a door at TechImage, it's possible that your original focus would return.

The next day, Malik Whitaker calls you at home and asks to set up a meeting. You agree to meet him at his TechImage office. At your meeting, Whitaker explains that Franklin is also talking with another prospective technical consultant. While this isn't a huge job per se, he assures you it will pay well. You ask discreetly whether Franklin and Whitaker would consider hiring Four Brothers for their needs in this matter.

"Well, we talked about going with a large-scale printer and maybe even just doing the easy route of conventional fire escape signs at all the exits, but we decided we wanted to go with an individual for a couple of reasons. First off, we're not sure we want signs that are completely conventional. We're trying to distinguish Dreamscape as somehow different from the other video rental places. We thought a creative individual might be better able to help us out with that," Whitaker said.

"But if you break convention completely, you run the risk that people won't recognize escape routes in a potential fire," you reply.

Whitaker nods. "That's why we want a creative professional from outside a larger business. We want someone who can come up with a recognizable image, but also make it classy and distinctive."

"Why can't a business like Four Brothers do that for you as well?"

Whitaker indicates a purpose for the freelance position here.

"Hosea and I have a thing about working in small circles. We like business at an individualized level. Even though Dreamscape is going to employ more people than TechImage, we fully anticipate being able to maintain a family atmosphere. This is especially important for the neighborhood we're working in," Whitaker tells you. "So we extend that smaller-scale working relationship as far as we can. Obviously, we'd work with a larger printer on some things, but for this sort of project we like to keep things to a one-to-one level. Anyway, the signs are only a small part of what we need here. We're after technical instructions that workers can readily access and understand, which is an order that most printers don't create on their own."

You talk for a while longer, and you agree to participate in a small "test," as Whitaker calls it. He offers you a layout of the Dreamscape floorplan and asks you to take it home, design one fire escape sign, and mark on the map where you believe the sign should be placed. Obviously, Whitaker acknowledges, the fire marshal would ultimately dictate where instructions and signs would be posted. Whitaker's purpose is to see what sort of sense for utility you have as well as your off-the-cuff design abilities. Your sign may be hand drawn if you choose, he tells you.

You have never blindly competed for a freelance job in this way and find the process and general lack of information Whitaker has offered somewhat eccentric. However, you have decided that the money and potential for future work is enticing enough to try for the job.

There are numerous unanswered details about the job, Dreamscape's needs, and Whitaker and Franklin's assessment criteria for the "test." While you could benefit from the experience, there may be some distinct drawbacks to vying for this freelance position. Discuss the benefits and drawbacks of the situation as you understand it to this point.

## Background Development

Consider the following questions as a way to fine-tune your understanding of the case and its details. Answer the questions alone or work in small groups of two or three to discuss the answers. Feel free to draw on your responses to inform any of the tasks that follow.

- The case raises an important question about conflict of interest. Discuss with your classmates whether you face an ethical dilemma in taking the freelance job with Whitaker and Franklin. Is the issue of conflict of interest changed if the freelance job develops into more technical writing for Four Brothers? What if this job develops into a permanent position at TechImage?
- Whitaker seems to have purposefully avoided giving you much insight, either into what he wants for this initial "test" or into Dreamscape and its overall safety needs. You must assume at this point that Whitaker is trying to be vague to see how creative you can be or how much initiative you might take. If you could craft a list of questions for Whitaker and Franklin regarding this job, what would they be? Make a list of the things you feel you don't yet know about this freelance job or about Dreamscape itself. Compare your questions with those of your classmates. How are they similar? Different?
- How are the demands for the "test" Whitaker has posed somewhat different from the demands you will face in the actual freelance job? What would you say is the focus, or the most important component for the job, particularly given the building's history? What do you believe Dreamscape's emphasis on safety should be? Compare your answers with those of your peers.
- Although Whitaker has told you to design some sort of fire escape sign, he has given you almost no parameters within which to work. Brainstorm about the different ways in which you could indicate a fire escape exit. Do all of the ways necessarily have to be one-dimensional signs? Do they have to use words you can read? What might be some alternatives to conventional signing? What risks do you take by deviating from a conventional approach?

## Gathering More Information

In teams of two or three travel to four or five different public spaces (e.g., libraries, hotels, businesses, hospitals, or schools) and examine the different types of technical instructions these places offer for fire escape or earthquake safety precautions. Make a list of their similarities and differences. Examine the contexts in which the instructions are offered. How visible are the instructions? What do they assume about the people they target (Do the directions assume literacy? Do they assume that all people in the building who may need to exit are sighted? etc.) Are such assumptions generally safe ones for the building and setting? How detailed are the instructions? How do the levels of detail vary according to place and need?

## Major Task 1

Design a fire escape safety tool that you would present to Whitaker and Franklin in a meeting. Also, to meet the goals Whitaker outlined for the "test," mark with an X on the map (see the case appendix) the location where you could see your sign working. Feel free to be creative with your design, but remember that it must serve its function—to guide people to safety in case of fire. Whitaker indicated that the price would not be of much concern, so as far as you know your design may employ whatever technology you see fit. Be prepared to present this idea formally and have a rationale for the design/information choices you have made.

At a later meeting, you present your idea(s), and Whitaker is extremely enthusiastic. Franklin is cooler but receptive. You are given the impression that Franklin has, from the beginning, backed the other prospective freelancer. However, it also becomes clear that your competition has not impressed either Franklin or Whitaker as much as you have. They offer you the job.

The most important component to the project, Whitaker and Franklin tell you, is the technical instructions you create for the employees at Dreamscape. They will be in charge of ensuring that the customers are safely out of the building in case of fire. Therefore, the instructions you create will serve as a reference sheet (probably to be posted in employee-only areas); however, they will also be used in the training manual for employees. Franklin and Whitaker assure you that the manager and assistant managers of both the video store and Entertainment and Eats Nook will "test" the instructions and drill new employees on procedures before Dreamscape opens. Their main problem at the moment is that they cannot allow you into Dreamscape right now because of the nature of the renovations taking place in the building. Your only choice is to use the relatively crude map they have provided. They have assured you that they understand your instructions may be subject to revision once you're able to get into the building and once the fire marshal reviews them.

Here are their stipulations for the instructions:

- There should be five "zones" corresponding to Dreamscape areas and exits. One employee for each zone will be scheduled at all times and will automatically assume the responsibility for customers in that area.
- Zone procedures need not be identical, but they should be similar so that employees who work a variety of zones will not have to memorize numerous procedures.

- Safety instructions should include checking all areas of the designated zone before leaving the building—*but only to the extent that the employee's life is not at risk.*
- The instructions should stipulate that under no circumstances should employees attempt to douse the fire themselves.
- Instructions should remind employees that they are responsible for keeping exits clear and doors unlocked in their designated zones.

## Major Task 2

Using the map and the stipulations Whitaker and Franklin have given you, create a set of instructions and design five zones that correspond with the exits at Dreamscape. Be prepared to justify your configuration of the zones orally.

## Correspondence Task

Write a memo to the state inspector, Janice Buchanan, summarizing the new set of instructions and signing you have created for Dreamscape. You may assume that this summary will serve as a precursor to Buchanan's on-site visit and inspection; she will not have visited Dreamscape prior to reading the memo. A copy of the memo should also be forwrded to Malik Whitaker and Hosea Franklin for their files. The memo should aim to summarize the techncial instructions, not entirely illustrate them, and also offer brief explanation for their purpose.

## Follow-up Task

It is several months after you have completed your project for Dreamscape. Throughout the project, you made it clear to both Whitaker and Franklin that you would be interested in a permanent position at TechImage if anything ever came up. You gave them your résumé and asked them to keep you in mind. Whitaker has just called you and told you that an entry-level position at TechImage will be opening in the next two weeks. He gives you some background on the position and asks you to send a cover letter to accompany the résumé he already has.

At the end of the conversation, he mentions that they have just hired two new employees at Dreamscape, both of whom have special needs. One employee, a video store floor worker and customer service assistant, is paraplegic and uses an automated wheelchair at work. The other is partially hearing impaired and works in the kitchen at the Entertainment and Eats Nook. Whitaker wonders if you might reexamine the instructions you wrote and evaluate whether they need to be revised at all to address either or both of these special needs. He asks you to drop him an e-mail or a memo on it once you've had a chance to review the materials. He makes no mention of payment.

Evaluate your response to Whitaker's request and write a memo to him with your answer.

## Case Appendix: Floor Plans for the Dreamscape Building

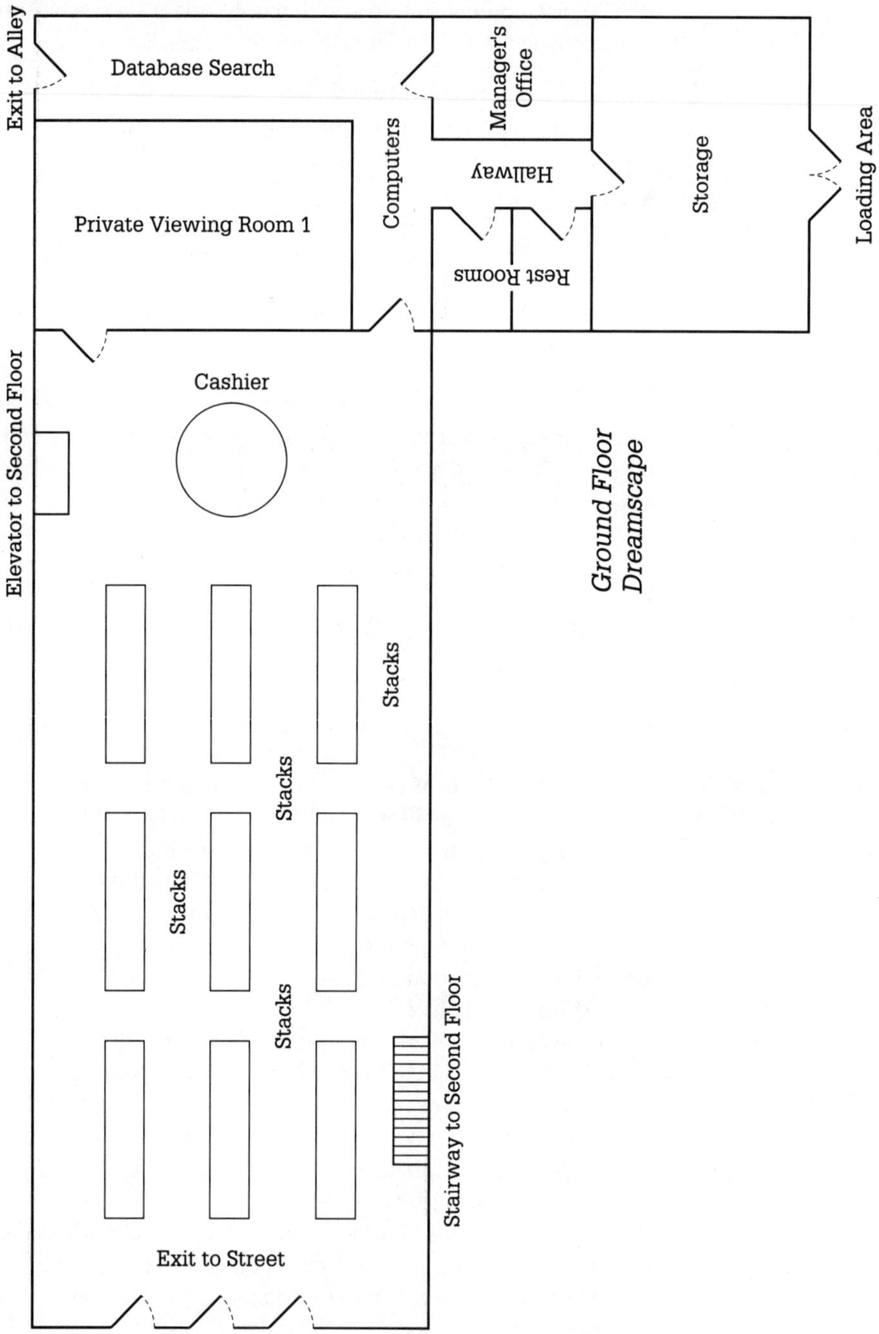

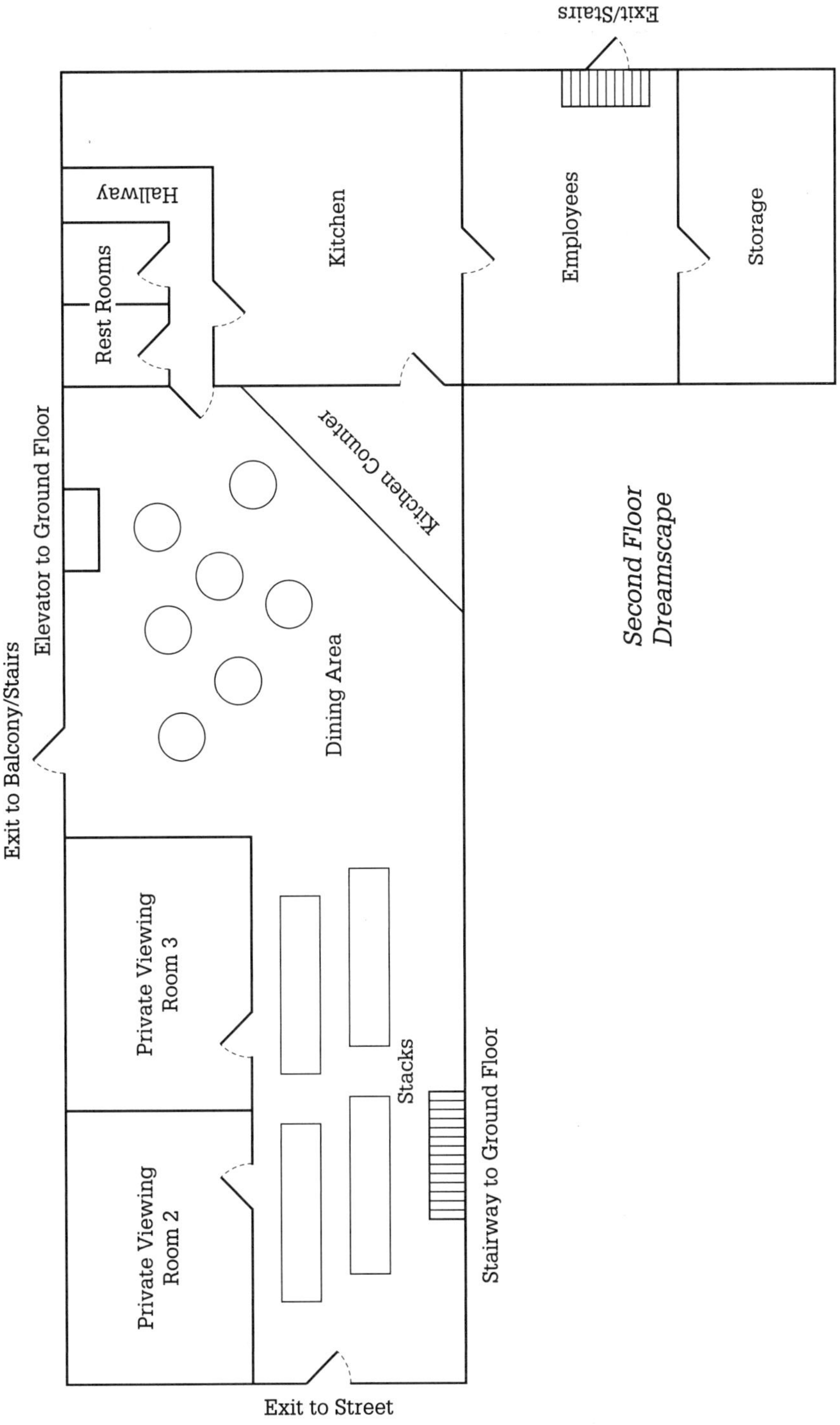

Exit/Stairs
Hallway
Kitchen
Employees
Storage
Rest Rooms
Elevator to Ground Floor
Kitchen Counter
Second Floor
Dreamscape
Exit to Balcony/Stairs
Dining Area
Private Viewing Room 3
Stacks
Stairway to Ground Floor
Private Viewing Room 2
Exit to Street

# Summary of Evaluative Criteria for Major Tasks in Case 14

| | *1 Unacceptable*<br>Insufficient answer to assignment expectations | *2 Below Average*<br>Inappropriate or ineffective verbal/ visual choices limit document success | *3 Meets Task Expectations*<br>Has answered objectives of assignment, but individual components could be strengthened | *4 Above Average*<br>Few flaws, document meets expectations, but could benefit from more attention to detail | *5 Excellent/ Professional*<br>Few or no flaws, demonstrates keen insight into case subtleties and details |
|---|---|---|---|---|---|
| **Purpose/Key Points**<br>▪ Identifies and meets purpose<br>▪ Demonstrates careful analysis of Dreamscape and its needs<br>▪ Demonstrates appropriate use of technical language skills and design elements | | | | | |
| **Context**<br>▪ Identifies/defines situational constraints<br>▪ Demonstrates understanding of ethical issues surrounding case demands | | | | | |
| **Audience**<br>▪ Identifies/defines audience and meets identifiable needs<br>▪ Establishes appropriate tone | | | | | |
| **Design**<br>▪ Demonstrates awareness of visual design elements of task<br>▪ Demonstrates an awareness of design options & technological aids in the development of these options<br>▪ Employs creativity when designing response to task<br>▪ Demonstrates an awareness of conventions for safety instructions | | | | | |

CASE

15

# *Haley-Grimes Corporation and the New Puree Pump*

In this case you are asked to write a report to supervisors about the testing of a new puree pump for a baby food manufacturing firm. Though the pump is the result of an internal proposal, it is not without its problems. You must negotiate differing team members' opinions as well as complicating contextual issues.

## Background

Note the introduction of a new line of Fancychild products.

Haley-Grimes Corporation is a large conglomerate that manufactures, among other food products, one of the top-selling baby food labels in the world. Fancychild Baby Products produces infant formulas and cereals; strained meats, fruits, and vegetables for babies; and an older child line including children's vitamins, vitamin-enriched toddler dinners, and teething pretzels. Fancychild has historically enjoyed a nationwide reputation for high-quality products at a reasonable market price. Because of its solid consumer satisfaction base, Fancychild has been able to expand to include a new line of organically grown and preservative-free line of strained baby food products, which sales forecasts suggest will be vastly popular.

Your role is defined here. Qualilty control officers are generally responsible for ensuring safety (both for the product and the work environment) and for overseeing standards of quality throughout the organization.

Haley-Grimes has 16 factories in the United States, nine of which house Fancychild manufacturing units. You are a quality control officer in a Fancychild division of Haley-Grimes located just outside Cleveland, Ohio. The Fancychild division where you are employed has 347 full-time workers, including 260 union line workers, 20 technicians (including quality control specialists), 25 engineers, 21 supervisory personnel, and 21 support staff personnel, (including secretaries, security officers).

## The Situation

Until recently, no division at Haley-Grimes had ever come close to striking. Haley-Grimes has historically prided itself on its excellent track record in collective bargaining with union officials. Three months ago, however, union workers at the Haley-Grimes in Indianapolis, Indiana, came very close to an eleventh-hour strike for better wages and job security. Just before workers went on strike, union officials and Haley-Grimes executives came to an agreement. While line workers at Fancychild in Cleveland do not appear to be dissatisfied with working conditions at the plant, everyone knows that all eyes have been on the Indianapolis plant. Haley-Grimes board members and shareholders uniformly acknowledge that strike is certainly possible anywhere, and they want to avoid such a possibility in whatever proactive ways they can.

In response to recent conflict, Haley-Grimes creates a proactive program. Note the purpose of the Leadership Action Program.

Therefore, Haley-Grimes executives have instituted a Leadership Action Program among union workers that encourages proactive proposals designed by line worker and engineering teams to improve working conditions and strengthen employee relations. The program encourages worker reinvestment in the organization and vice versa. Line workers feel as though their ideas are valued, and the entire corporation benefits from innovative plans to enhance working conditions. Each of the 16 Haley-Grimes plants offers $250,000 in grant money to be divided among recipients with the most attractive proposals. In addition to the satisfaction teams receive from their role in improving the workplace environment, team members also receive a bonus package at the end of the fiscal year.

One of the first proposals accepted and funded for the Leadership Action Project was a plan produced by a team of six line workers and two engineers at your plant in Cleveland. The line workers operate a main pump for the fruit puree processing division, and their proposal aimed to alleviate some of the inefficiencies they perceived with the pump as they used it on the new organic food line. While Haley-Grimes manufactures the puree used in much of its infant foods, it is stored initially in 50-gallon barrels and shipped from a main warehouse out to the various plants. This is in part because each of the plants is designated to produce and package certain foods, all of which use the puree differently. After the barrels of puree are shipped to the plant, line workers prepare and package the food appropriately. Currently that process involves operators aligning 50 gallon barrels of puree mixture with a pump. The pump then extracts the puree and sends it to the packaging line where it is put into individual plastic jars for shipping. The pump looks like the one illustrated on the following page.

What problem did the proposal from the Cleveland team try to address?

The proposal argued that because the puree in the new organic foods line was a different consistency from the other puree used in the traditional strained fruits and vegetables, the current pump was not as efficient as it could be. Part of the problem, they reasoned, was that the suction cups and hoses were not large enough to handle the new, grainy mixture. As a result, hoses would sometimes become clogged and have to be removed and flushed. In addition, the barrels were somewhat awkward and required changing every 10 to 15 minutes. The process of taking the barrel out and installing a new one required two line workers and at least 7 minutes.

Consider the benefits of the proposed use of the plastic-lined boxes over

Together with two mechanical engineers, the line workers designed an alternative pump that theoretically would be more effective and time efficient. In

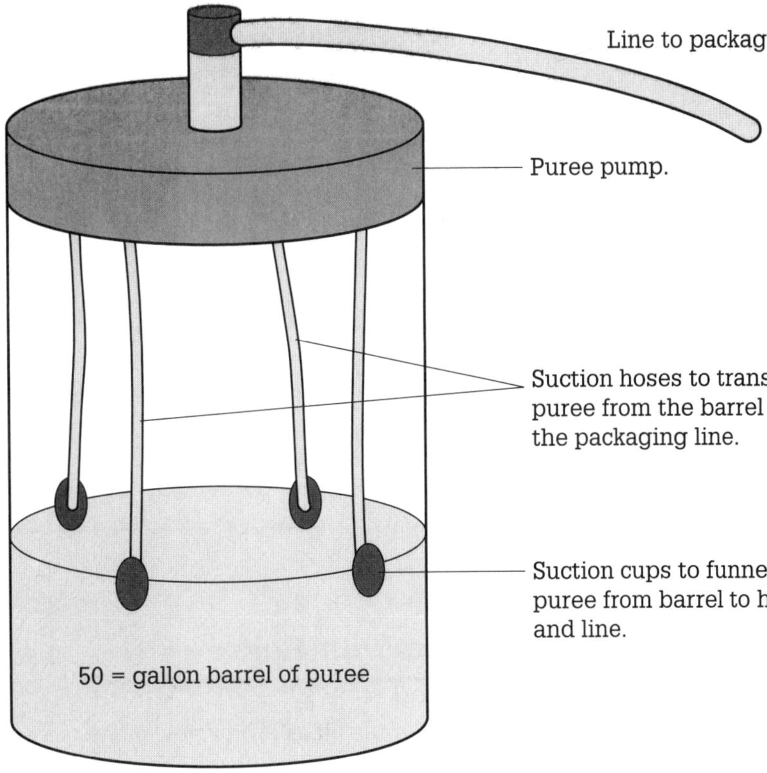


the metal barrels. What are the benefits to workers?

What does it mean that the Cleveland plant will serve as a test case for the Leadership Action Program?

addition, the team had researched alternative storage packaging methods and located a manufacturer that guaranteed a plastic-lined cardboard box would fit Fancychild's needs better than the current 50-gallon barrels. Not only was the cardboard box more environmentally sound, but it could hold 60 gallons of puree, which meant less time changing containers after one was empty. The engineers designed the new pump to operate with the cardboard boxes. The result was the design illustrated on the following page.

The team offered the proposal to Haley-Grimes's board of directors, and all were favorably impressed. The proposal argued that if the plan were implemented, employees would benefit by being able to pay more attention to other pressing issues on the line such as safety and communication, while the company would benefit by saving money in the long-run with improved efficiency. Therefore, the board of directors approved the proposal and hired a Denver plant at $15,000 to construct the pump the proposal team had designed. Haley-Grimes board members agreed to use the Cleveland plant as a test case, and if the pump proved efficient, the board would approve a switch in all plants. The Cleveland team's proposal was, by all accounts, a superb example of what the Leadership Action Program was designed to reward—initiative, creativity, and loyalty to the organization.

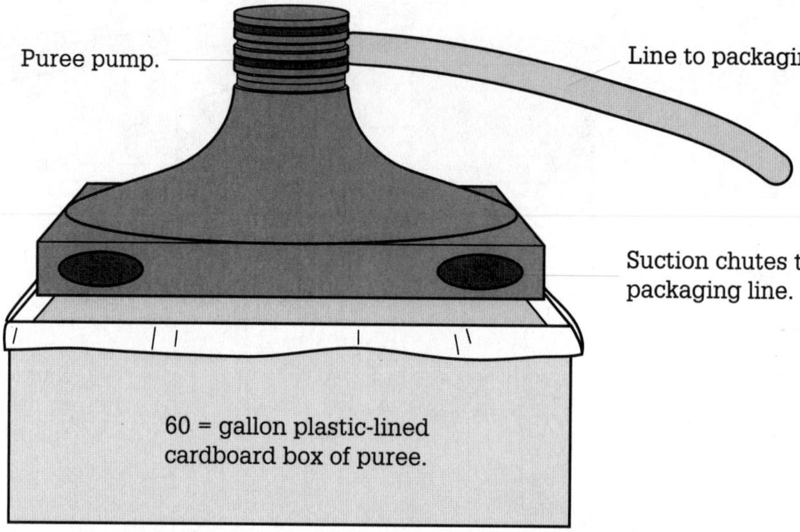


## The Task and Your Role

The purpose for your team is briefly defined here.

It is two months after the plans for the new pump were sent to the Denver manufacturer. The Cleveland plant manager, Craig Haute, has formed a team of four to fly to Denver for three days to test the newly developed pump.

The team has been asked to submit a report to the board of directors summarizing your findings. Just before you leave for Denver, you learn two things from your supervisor. First, the purchasing office has made a monumental error by ordering only boxes of the puree. This means that once the line runs out of barrels of puree (you have approximately three weeks' worth of barrels in stock), your only choice is to use the boxes or shut down until you can get more barrels (in four to six weeks). A shutdown would cost the company a great deal of money and is not desirable. Second, the board of directors wants this project to be successful in order to use it as a positive example nationwide. It sends the message that all employees have influence and power, a message that is especially needed in places like Indianapolis. The subtext to your discussion with your supervisor appears to be, "Don't make waves."

Note the problems you discover with the pump. How are they compounded by circumstances offered in the preceding paragraph?

After testing the pump, your team learns that it has several problems. First, the cardboard boxes tend to buckle easily under the pressure of the pump. The buckling slows down the pumping mechanism. This problem can probably be solved, at least in part, by building metal side supports that would fit around the box when it is inserted under the pump. Second, as Goldstein and Kovaciavek changed boxes, they appeared to be saving no time—the process still took approximately 7 to 10 minutes. They both pointed out, however, that part of this was because they were unfamiliar with the feel of the boxes and the pump. They agreed that the more experience workers had with the machine,

| | |
|---|---|
| **Team Member:** | **You** |
| Role: | Quality control |
| History: | You have been with Haley-Grimes for only two years. You know and respect your team members, despite the fact that you do not always agree with them. |
| Investment: | Moderate—while your job does not necessarily depend on this situation, you are still proving your worth to Haley-Grimes. |
| **Team Member:** | **Elliot Goldstein** |
| Role: | Line worker |
| History: | Goldstein has been with Haley-Grimes for 27 years and was a part of the team that proposed the new pump. Goldstein is widely respected by fellow line workers and is an extremely influential union member. |
| Investment: | High—Goldstein is heavily invested in the success of this pump because it is essentially his brainchild. His job does not depend upon its success, however, so the investment is largely emotional. |
| **Team Member:** | **Anna Kovaciavek** |
| Role: | Line worker |
| History: | Kovaciavek has been with the company for only six months and was not a part of the proposal team. However, her job is to run the pump and understand it. A Croatian immigrant, Kovaciavek is a nonnative English speaker. |
| Investment: | Moderate to high—while the plan was not hers, Kovaciavek is invested in seeing the pump work well. |
| **Team Member:** | **Pat Locker** |
| Role: | Mechanical engineer |
| History: | Locker was a part of the team that proposed the pump and has been with the company three years. Locker and Goldstein are close friends, and Locker is easily influenced by Goldstein's opinions. |
| Investment: | High—again, Locker helped to put the proposal together. In addition, a success or failure might partially influence Locker's longevity with the company. |

the faster they would become. Finally, some last-minute changes in the design of the base of the pump and some scheduling problems with shipping have led to a slight delay. The Denver manufacturer guarantees that the pump will be installed in 15 days; this is almost a full week later than the original installation date. Normally, this difference would mean little except for the dwindling number of barrels in your warehouse.

## Background Development

The following list of questions may help you to sort through the dynamics of this case and the issues it deals with. In a group including three of your peers, consider the questions as they pertain to the information available to you in this case. Write your answers down and feel free to use them as you develop responses to any of the following tasks.

1. It is clear that the Haley-Grimes team faces several important purposes in this case. List the purposes you can define from the information provided in the case and rank the list in order of importance.
2. How would you define the limitations or the problems the team faces in this case? Briefly list and rank the problems in the order you perceive to be most pressing or most important first.
3. Discuss with your group how you would assess a pump as the one described in this case. Is the team offered any sort of criteria by which to judge the machinery? What criteria might you use for such an assessment? If you can name criteria you believe would be useful in judging the new pump, make a list and rank the points in order or most-to-least important.
4. Using the list of team members, consider the weight of each team member's opinion on the worth of the new pump. While Anna Kovaciavek acknowledges that the pump cannot *harm* the current process, she is also not convinced the new pump will *help* things, either. Discuss with your team members the potential problems in disregarding Kovaciavek's misgivings. Do these potential problems outweigh the alternative problems of not accepting the pump as it currently stands?

## Gathering More Information

Your team has discussed matters thoroughly, and you have decided that to address management's expectations directly you need further insights into what the evaluative criteria for the pump must be. In acquiring these criteria, you also may take some of the pressure off your team for making this decision on its own. Write a memo to the board of directors requesting the criteria by which you should judge the pump and briefly acknowledging some of the potential problems you see. The board of directors chair is Duane Bowers, and the address is Haley-Grimes Corp., Fancychild Division, 1209 Summerset Way, Detroit, Michigan 86521. You should copy this memo to your plant manager in Cleveland as well.

## Outlining Criteria

Duane Bowers, board of directors chair, has faxed you his response to your request for clearly defined criteria. Bowers suggests that the criteria should be defined by the team and, because you are the quality control officer, that it is probably your job to lead in those decisions. He notes that he would expect the standard criteria for evaluation to be used (e.g., efficiency, productivity, cleanliness, and the ease with which the new machinery may be used), but you may develop those criteria as you see fit.

The main message in Bowers' memo was to take control over the assessment and simply do it, which your team agrees doesn't help you much. So you sit down with members and try to define the criteria by which you will judge the new pump. Together with three others (each assuming a role in the case), outline five criteria you believe will be helpful to use in your assessment and create a scale by which you can judge them. Then, as a team, discuss how the pump fits these criteria.

## Major Task

Because you are the quality control officer, you are designated as the writer of the report your team will submit to the board. Your report should be in letter form and addressed to the board of directors. It should reflect your assessment of the pump. While Locker and Goldstein adamantly argue that this pump will better serve the organization, Kovaciavek is less convinced it makes much of a difference. Kovaciavek, however, acknowledges that it cannot technically *harm* the extracting process as it currently stands. Using the criteria the team has outlined, you have agreed to write a draft of the report, which the team will review and respond to the following afternoon.

## Team Review

In a group with three other people (each assuming an appropriate role from the case), review the draft together and discuss revision.

## Follow-up Task

While your team certainly discovered some problems with the pump, you believe that the Leadership Action Program offers enormous benefits for Haley-Grimes and its employees. At the next board meeting, you and Elliot Goldstein have been asked to come and give a brief presentation on your team experience in Denver. You have decided to focus your own part of the presentation on what you see as the benefits of the Leadership Action Program and, in particular, how you saw evidence of those benefits in your experience on the testing team. Outline and present a 10-minute oral presentation on your experience and your assessment of the Leadership Action program.

# Summary of Evaluative Criteria for Major Tasks in Case 15

| | **1 *Unacceptable***<br>Insufficient answer to assignment expectations | **2 *Below Average***<br>Inappropriate or ineffective verbal/ visual choices limit document success | **3 *Meets Task Expectations***<br>Has answered objectives of assign- ment, but individual components could be strengthened | **4 *Above Average***<br>Few flaws, document meets expectations, but could benefit from more attention to detail | **5 *Excellent/ Professional***<br>Few or no flaws, demonstrates keen insight into case subtleties and details |
|---|---|---|---|---|---|
| **Purpose/Key Points**<br>▪ Identifies and defines purpose<br>▪ Articulates key points clearly<br>▪ Demonstrates an ability to analyze potential evaluative criteria<br>▪ Works easily with others and with attention to group dynamics | | | | | |
| **Context**<br>▪ Identifies/defines context and situa- tional constraints<br>▪ Demonstrates awareness of docu- ment situatedness | | | | | |
| **Audience**<br>▪ Identifies/defines audience<br>▪ Establishes appropriate tone<br>▪ Understands technical details enough to commu- nicate effectively | | | | | |
| **Organization**<br>▪ Demonstrates ana- lytical insights<br>▪ Develops a clear line of argument<br>▪ Employs identifiable, appropriate pattern of organization | | | | | |
| **Oral**<br>▪ Summarizes effec- tively<br>▪ Demonstrates confi- dence with informa- tion and delivery | | | | | |
| **Design**<br>▪ Demonstrates awareness of visual design issues in analysis | | | | | |

CASE 16

# Meeting Sitka's Medical Needs: A Proposal for Updated Technology

In an effort to stay competitive and retain doctors, a small hospital in Alaska has determined that it must update its technology in order to increase efficiency and diagnostic capabilities. Writers are asked to serve as development specialists to research prospective foundations and write a proposal for new computer equipment or telecommunication capabilities for connecting with larger research institutions.

## Background

Locate Sitka on a map and notice its geographic remoteness.

Sitka is a small island located off the extreme southwest coast of Alaska near Juneau. Unlike the common conceptions of harsh Alaskan weather, Sitka's island climate is mild, rarely dipping below zero in the winter and rarely rising above the 80s in the summer. With a population of approximately 9,000, Sitka is the state's fourth largest city. It supports one small private college as well as a U.S. Coast Guard station and a relatively stable business district. Census data suggests that the community is growing and is economically more stable than it has ever been.

As small as the community is, it does support two hospitals, in part because Sitka is geographically isolated from the mainland and from any significant population base. As a result, Sitka is required to remain self-sufficient in its health care services or risk lives by having to fly critical cases out to Seattle, which is 60–90 minutes by air.[1] One of the ongoing issues facing Sitka—one that is not unusual for many small, geographically isolated communities—is its ability to retain doctors in its hospitals. Because of Sitka's location, it is often difficult to hire and keep qualified physicians and some of the necessary specialists to consistently meet all of the community's medical needs.

[1]While the background information on the community of Sitka is accurate, the history of the hospital highlighted in this case is fictionalized and based, in part, on a case study of Woods Memorial Hospital in Etowah, Tennessee from an article by Joshua Macht, "Critical Care," in *Inc. Technology*, no. 2 (July 1996), pp. 61–65. Sitka, Etowah, and other geographically isolated/low population communities face serious critical care issues as federal aid declines. More and more rural hospitals will be obliged to turn to private foundations to strengthen quality of care and efficiency, or communities may lose the health care facilities they have.

Think about hospitals you are familiar with. Are there departments or services you see missing from Carney?

The older of the hospitals, Carney Medical Center, was established in 1948 and currently offers 90 beds; emergency room facilities; intensive care; three operating rooms; outpatient facilities; X ray, radiology, and diagnostics; obstetrics and pediatrics (housed in adjoining wings); a small cancer ward; counseling and mental health services; and a pharmacy. Most administrative processes are still carried out manually at Carney. For example, in patient billing, clerks have historically calculated the various drug, supplies, and procedures charges for each patient according to a list (often consisting of handwritten notes and pages filled with actual price tags from the supplies used) compiled by a variety of nurses from different shifts. The billing clerks then calculated charges and manually typed insurance forms, sending duplicate copies to the patient, insurance agency, and hospital billing administration. The process has been time-consuming, inexact, and costly, but it is also not unusual for a small operation such as the one at Carney.

Carney completed a self-study three years ago, which suggests that the board of directors was aware that Carney's systems and personnel needed to be examined.

Three years ago, Carney's board of directors completed a year-long self-study of the hospital's operating systems. It found that despite a strong staff of medical professionals, there were several major issues Carney needed to address immediately. First, Carney was losing money. While this was partly the result of changes in Medicare coverage and federal assistance, many of the losses could also be blamed on inefficiencies in time management and antiquated record-keeping practices. What the study showed, however, was that Carney was only bringing in, on average, 72 percent coverage of total Medicare and costs. In addition, Carney was also experiencing a higher percentage of patients with no insurance coverage and an inability to pay for services. Second, in the last five years, Carney had lost three of its residents to other hospitals in Anchorage, Alaska; Seattle, Washington; and Eagle Grove, Wisconsin.

Ami Berger's experience exemplifies a larger problem for new doctors and residents at Carney. What is it?

The most promising of the young residents, Dr. Ami Berger—who had only been with Carney for 18 months before her departure—cited the loss of a patient for the reason she needed to leave Sitka. An 8-year-old Sitka native was admitted with abdominal pains and fever, which Berger diagnosed initially as appendicitis. When the surgical team found no evidence of infection or cyst, Berger sought another reason for the obvious infection while treating the child with antibiotics. Limited by a lack of adequate resources and concerned with her patient's deterioration, Berger determined to fly the patient to Seattle, where the larger research hospital would be better equipped to diagnose and treat the young patient's acute condition. The patient arrived safely but died in Seattle 24 hours later of hemorrhaging caused by a rare allergic reaction. Though no blame or negligence was ever assigned to Berger, the doctor believed that had she possessed better diagnostic capabilities at Carney, precious time would not have been lost in transporting the patient to Seattle. Although the other residents who left cited personal reasons, including a need to be closer to family or to live in a more metropolitan area, the hospital's self-study suggested that Carney offered few incentives for young doctors to work there, in part because its technology and access to resources was not comparable to the facilities most of the doctors who came to Carney had trained on.

After some shuffling of administrative posts directly following the self-study's release, Carney's board of directors determined that it was essential to hire a new hospital administrator with a different vision for Carney. If Carney hoped to stay afloat into the next century, one board member argued, it needed a leader with a "progressive health care vision." After conducting a search throughout Canada and the United States, the board finally hired Al Willig from Minneapolis, Minnesota. Willig had extensive experience at the Mayo Clinic as well as the Harvard School of Business where he was trained. Carney's board of directors was attracted to Willig's clear understanding of the bottom-line quality issues Sitka's small hospital faced. Initially, hospital staff were quietly skeptical about Willig because he had no experience with such a small community hospital and he was not an Alaska native. However, Willig quickly established himself as a passionate, dedicated advocate of quality health care, and the staff eventually acknowledged a grudging respect for the outsider.

## The Situation and Your Role

About six months after Willig arrived in Sitka and began his own personal examination of Carney's operations, he went to the board of directors and argued successfully to hire a development specialist. While part of Willig's own job description noted that he was in charge of all fundraising activities for the hospital, he pointed out that this was an inefficient use of his time, given all of Carney's restructuring needs. Instead, Willig argued, he needed someone he could trust to research and effectively reach national and international health care foundations and individual donors in order to boost Carney's assets. Such a task was a full-time endeavor, and he guaranteed the board that within two years after hiring a development specialist, Carney would see significant results and changes in its systems. While the board was initially reluctant given that the hospital was currently losing money, they eventually agreed that in order to turn Carney around, they would have to spend a little money. However, they could not afford much of a salary, so Willig would have to work hard to find a competent development officer to do the job for less than the going rate.

Your role is defined here. What do you anticipate doing in this role?

Willig hired you as his development officer because of your skills in research and writing abilities and because you had temporarily located to Alaska for personal reasons. Your job is to help Willig assess the specific needs of the hospital, research possible foundations and private donors, create successful proposals, and establish positive hospital-community relations to cultivate donors for matching-grant opportunities. Your first order of business is to read through the self-study, concentrating on its Needs section, and identify the top five goals for Carney.

"So what would you say Carney's top five needs are?" Willig asks you in your Wednesday morning meeting during your first week on the job.

You glance down at your notes and briefly explain what you have come up with. "First of all, I don't think Carney is unusual in its need to increase retention among its professional staff. The staff we have here now is extremely competent, but based on this report, there is high turnover in the most recent hires over the last five years. So the first goal is to increase retention."

Willig nods, and holds up his hand for you to stop. "And what do you think will do this?"

"Increase retention?"

He nods.

You look at your notes again. "The self-study suggests that the biggest complaint among professional staff is a feeling of isolation and frustrations with the limited resources. I think the new hires need to have access to the latest technology that most of them used at the universities or medical schools before they got here. I was just reading an article on what advanced computer networking can do to increase efficiencies in hospitals just like this one. And from my experience at the university, I know more and more school and information systems are hooking up with remote areas such as Sitka with telecommunications networks to share conferencing capabilities. I think those sorts of things would help new staff to feel as though they're not quite so isolated. "

Your purpose is defined here. What are the questions Willig wants to have answered?

Willig nods again and sighs. "You're right, and this is where I want to get started. It's pretty clear that the community and the board are not able to support the technological advancements we're talking about, so we have to seek out the sources who are. I think our first task is to find out what is available in terms of money for technology startup systems. I'm specifically interested in computer networking and telecommunication/fiber optics options. I want to know how much the stuff is going to cost and who we can get the money from. Then we'll sit down and think through how to approach them."

Willig seems to be dismissing you, and you are a bit confused as you look at your notebook. You still have four more points to make. Willig smiles as he leans forward. "I figure you have a couple of weeks worth of work to do here. We'll get to your other four points once we get close to finishing this one. As I see it, your points are the summary of what you're going to be doing this first year in Sitka." It is clear that he was testing you to see how carefully you had read and assessed Carney's needs.

"OK," you nod. "So what's the first step? I'm not even really sure what sort of technology we're interested in employing here. There are a lot of needs where the medical staff and the hourly staff are concerned. Are we looking at the computer networking for both?"

"At this point, nothing is networked and our best bet is to overshoot first. So, yes, I think we're looking at a project that would fund and wire the whole hospital. It won't be cheap, but it's only the first step in bringing the hospital up to speed. You mentioned only computer networking. I have a friend at a hospital in Tennessee, and he was telling me about everything they've done to coordinate networked computers *and* telecommunications—television monitors, cameras, and satellite links—to allow doctors to consult with specialists at other hospitals. The more I think about the possibilities of doing stuff like that here, the more I become convinced we won't be losing folks like Ami Berger because she's discouraged at losing a patient to primitive resources."

This paragraph points out a slight difference in understanding of your purpose between you and Willig.

You nod. "I agree with you in principle. These are resources that ensure Carney's future success. But—you hired me to be the pragmatist, remember—I'm skeptical about finding a source that'll do all of that for us. We're talking about needing to raise an endowment here, and we don't have that yet. I suggest we shoot a little under the line at first. I don't want to start out with a failure. The board, as you well know, doesn't want to see that."

"You're probably right," Willig agrees. "I don't want to start out with a failure. Do some research and see what you can come up with in terms of foundations and information about what we need." He laughs, "Assume we're starting from ground zero. We have nothing and we want everything. See where that takes you."

Your task requires that you begin research on the kind of networking systems appropriate for Carney's needs. As you begin to look into systems, pay no attention to cost as a factor for considering what will be appropriate. Look only for what technology would be best for Carney.

## Background Development

Consider the following questions as a way to fine-tune your understanding of the case and its details. Answer the questions alone or work in small groups of two or three to discuss the answers. Feel free to draw on your responses to inform any of the tasks that follow.

- Reread the background and situation sections of this case. Consider your purposes, and list them briefly on a piece of paper. Now rank-order them in terms of importance.
- A large part of this case requires you to do background research. Some research may involve talking with experts; some may involve library time or time spent on the Internet. But before you begin research, you must have questions you need to have answered. Make a list of all the questions that come to mind after reading the background to the case. Don't worry about how huge the questions may seem—just list them out. After you've finished making your list, reorganize the questions so that they reflect an order in which you can pursue the answers. Or you may choose instead to list them according to their importance. Finally, sit down with a partner and exchange your lists. Examine how different and/or similar the lists are. Are the questions focused enough? Can you narrow them or make them any more specific before you begin searching for answers?
- A wonderful way to find information (like how many computers and what sort of networking system a medical facility of 90 beds might need) is to do a search on the Internet. Another way might be to talk with representatives/administrators at local hospitals to determine what they have and/or what they feel they need in order to operate more efficiently. Choose a partner and divide your tasks. One of you may surf the Net, while the other may choose to interview local administrators or do a more traditional library search. Once you have finished a preliminary search, compare notes with your partner. What have you learned about networking a large organization?
- It is sometimes helpful to read about similar cases in order to fine-tune your understanding of the situation. Locate in your library the July 1996 issue of *Inc. Technology* and read the article by Joshua Macht entitled "Critical Care." Discuss in a small group how the situation in Etowah, Tennessee is similar and different from the situation in Sitka. Is there anything in the Etowah case that might be helpful to you as you develop your response?

## Gathering Information

Research the costs of installing a computer networks system that would link approximately 35 personal computers with large-screen monitors (e.g., Power Macintosh 7100/80s or Gateway 2000 Crystal Scan 14s), 15 laser printers, and a server that would connect them all to the Internet. In addition to the purchase of hardware, software and network links, you should also consider the cost of training a network manager (this person would already be an employee who could manage the network for the hospital in addition to his or her other duties). You may contact computer specialists, local hospitals, and texts and journals to locate your information. It is also possible that you have a computer specialist on your campus who would know some of the information you seek or who could direct you to an appropriate source. Seek that person or persons out and explain your purpose to them. Keep good notes on the sources you use so that you can eventually attribute the information you've gained. Once you have located information you believe will be useful to Carney, summarize it briefly in a journal for your own use.

## Major Task 1[2]

Using the information you have gathered, write a memo to Al Willig explaining what you have accomplished. For his benefit, create a chart that shows three possible sources that could outfit Carney with networking capabilities, hardware, appropriate software packages. Be sure to also include estimated costs. This chart should help Willig better understand what is out there and what you will face as you attempt to advance Carney's technology.

## Major Task 2

Using a book of grant foundations and private sponsors (found in most libraries), examine Carney's options for pursuing private money to support advancing the technology through a computer network system. Write a short report to Willig that summarizes the foundations you believe may be interested in funding Carney's Technology Enhancement Project. In this report you may suggest which of the foundations looks most promising and why. Write this report in letter form.

## Formal Correspondence

While many foundations actually send out formal requests for proposals, they are unlikely to be listed in their entirety in the grants books you'll find in your library. What you *will* find, however, are contact names and addresses to which you may write for more information. Choose one of the foundations you believe looks promising for Carney's Technology Enhancement Project and write a let-

[2] Teachers may choose to ask students to investigate telecommunication options as opposed to computer package options for the tasks.

ter requesting information about the foundation's formal application process. In the letter, briefly explain who you represent and what your purpose is (because this is a letter requesting information, you do not need to go into much detail).

## Major Task 3

Write a proposal to Carney's board of directors that outlines your rationale for pursuing the foundation(s) you have targeted as good prospects for Carney's Technology Enhancement Project. In the proposal to the board (who must approve any formal grant project), you must outline the *need* for technology enhancement that you and Willig have discussed. Assume that you will attach a copy of the Carney annual budget with your proposal. The address is Carney Memorial Hospital, 1108 Plainview Road, Room 100, Sitka, Alaska 99601. Carney's board consists of the following members:

**Carney's Board of Directors**

| | |
|---|---|
| Barbara Teesloe, Chair | President, Citizens National Bank |
| A. J. Red Deer | Sitka School Superintendent |
| Janice Ewing | Podiatrist, Carney Memorial Hospital |
| Truman Weathers | Retired air force captain |
| Lucas Wagner | President, Sheldon-Jackson College |
| Smokey Guillaime | Nurse, Carney Memorial Hospital |
| Frances Blackwood | Owner, Blackwood Taxidermy |

## Major Task 4

Write a proposal to the foundation of your choice to fund all or part of your project to enhance Carney's computer technology. Follow the proposal guidelines set up by the foundation. Assume you will provide a copy of Carney's annual budget in the proposal as an appendix.

# Summary of Evaluative Criteria for Major Tasks in Case 16

| | ***1 Unacceptable*** Insufficient answer to assignment expectations | ***2 Below Average*** Inappropriate or ineffective verbal/ visual choices limit document success | ***3 Meets Task Expectations*** Has answered objectives of assignment, but individual components could be strengthened | ***4 Above Average*** Few flaws, document meets expectations, but could benefit from more attention to detail | ***5 Excellent/ Professional*** Few or no flaws, demonstrates keen insight into case subtleties and details |
|---|---|---|---|---|---|
| **Purpose/Key Points**<br>▪ Identifies and meets purpose<br>▪ Articulates key points clearly<br>▪ Demonstrates research ability and resourcefulness in locating information | | | | | |
| **Context**<br>▪ Identifies/defines context and situational constraints<br>▪ Demonstrates awareness of proposal and report(s) situatedness | | | | | |
| **Audience**<br>▪ Identifies/defines audience and meets identifiable needs<br>▪ Establishes appropriate tone<br>▪ Locates and analyzes technical informaltion<br>▪ Proposal offers complete, accurate information | | | | | |
| **Design**<br>▪ Demonstrates awareness of visual design elements of task<br>▪ Addresses issues of readability and accessibility when charting information for Willig | | | | | |

CASE 17

# *Architectural Risks: Lowe & Company's Dilemma*

Lowe & Company is a relatively new achitectural firm with a strong desire to land a local suspension walkway job. With little experience in suspension walkway engineering, Lowe & Company seeks to forge a collaboration with a firm that can help fill the experience gaps. This case asks you to write a series of letters and memos as elements in the longer-term negotiations of partnerships.

## Background

Consider the problems a brand new firm faces in getting started. Can you list some of the hurdles Lowe & Company might face?

Lowe & Company is a new architectural firm located in Milwaukee, Wisconsin. Struggling to get off the ground financially, Lowe has worked hard to offer reasonable, honest bids for potential clients in order to establish a name for the firm. To date, their competitive edge has been Lowe & Company's ability to accomplish relatively large projects for less money than the competition.

Lowe & Company was founded 26 months ago by architect Maggie Lowe and her brothers, Patrick and Shawn. Patrick is a mechanical engineer, and Shawn received his M.B.A. from the University of North Carolina last May. Together with six other young architects and four engineers, the three siblings have managed to land two substantial accounts and five short-term projects since they opened their modest office downtown. Clearly, competition in the area is stiff. The first substantial account involved designing a public space for a downtown redevelopment project. The public space included a brick walkway, a koi garden, landscaping, and an entry arch, as well as strategically placed public art commissioned by the city. While the work itself was not difficult, it *was* noticeable. It was because of this public work that the city council asked Lowe & Company to offer a bid to design and build a suspended walkway system for the ongoing downtown redevelopment project. The system would include five strategically placed suspended walkways between major structures in the main business district. Lowe & Company was the only small-scale architectural firm in the area asked to compete; to win the business would mean an enormous boost for the firm.

## The Situation

Your role is defined here. What would you anticipate your background and training to be?

You were hired three months ago as an engineer with Lowe & Company, and your work has been limited to the short-term projects involving home additions and landscape design. Maggie Lowe has called a special staff meeting to discuss the city project.

Maggie is expressing reservations about bidding for the city job here. Why?

"I have to admit," Maggie begins, after circulating the copies of the city's request for bids, "I feel a little out of my league here. I have never worked on a suspension project before, and the only experience I do have comes from what I learned in the classroom—and *that* was several years ago."

"Curt and I worked on that small suspension bridge project in Minnesota back in 1989, which will probably be helpful," offered Patrick. Curt Jackson, seated next to you, nods. Jackson was one of the original engineers hired by Lowe & Company because he and Patrick Lowe had worked together in Minnesota prior to the Lowes' partnership two years ago. Jackson is quiet but meticulous in his work, and he has already given you some helpful insights in your own work.

Maggie narrows the options for this project. Does she consider all of the possibilities? Discuss with your classmates.

Maggie shakes her head. "I remember that project, guys, and I respect the work you did. But I don't think one project between thirteen of us makes us qualified to do this. I think we have a couple of options. Tell me what you think of them. First, I wonder how committed the city is to a *suspended* walkway. Seems to me that's pretty limiting in terms of what they might consider. One option might be to offer an alternative, something we *do* feel comfortable designing."

Shawn let out an audible sigh. "Whew! I thought you were going to say that we ought to opt out of the competition. Frankly, based on the books right now, we really can't afford not to try on this one."

Note that Lowe & Company's financial situation suggests a constraint in their options. What is it?

Several heads around the table nodded as Maggie smiled ruefully. "Yeah, I'm well aware of our tight purse strings at the moment. At the same time, we certainly couldn't afford a major lawsuit if we designed something that failed because of our own inexperience. I don't want to be responsible for creating something like the suspended walkways that collapsed in Kansas City in 1981 because I was too inexperienced to see a flaw in the design detail or because I figured the safety factor wrong. Personally, I think the city has offered us the chance to bid only because we did a nice job on the public space earlier this year and we were cheap. I think it's unlikely we'd land this thing on our own, though. Too many liabilities."

"What do you mean, 'on our own,' Maggie?" Your interest is piqued.

"Well," she responds, "I've scanned this call for bids over and over again, and there is nothing that says we can't team up with another firm. Wait. Before you protest, Shawn, let me finish my line of argument here. Since we don't have any real experience with suspension—but we *do* have several things to our advantage—we might be likely candidates for the team approach. If we enlisted the experience of another firm to help construct a couple of viable models, we could not only draw on our own strengths in detailing and knowledge of the "larger picture" of the downtown redevelopment plans, we could also get some invaluable experience with suspension design."

"But if we team up with another firm, Maggie, it means we're bringing home less money," Shawn argued.

Maggie shook her head. "Not necessarily. I grant we might clear less from the outset, but imagine what landing this might do for business in the long run. It's a matter of weighing short-term versus long-term benefits."

Consider the tension between Maggie and Shawn. They are driven by different priorities. What are they?

Shawn was unconvinced. "If we don't meet our short-term obligations, Maggie, we're not going to have to worry about long-term benefits."

"But Shawn, I doubt we'd land this thing on our own anyway. Some money is better than none."

"I have a suggestion," Patrick said. "Why don't we divide up into teams and do a little research? We could come back with some reports first thing tomorrow morning and from them make a decision about our approach to the city." Heads around the table were nodding at this point. He continued, "Could one team look into possible partnership angles and the other devise a couple of alternative designs for the walkway?"

## The Task and Your Role

What skills are necessary to research the potential partnerships?

You are a part of the team that will research potential partnerships. This task is something you are confident with since you have no actual experience with bridgework. You are dubious about Lowe & Company's undertaking such a challenging project, despite the fact that you realize the firm is in need of the kind of windfall another city project would ensure. You will be working with Maggie, Curt Jackson and two architects, Calle Emerson and Aaron Matthews. Despite Curt's limited experience with suspended structures, Maggie and Patrick both agreed that Curt should be tapped because of his contacts with the Minnesota firm, Urbana Design, at which he and Patrick were employed for several years.

Here your purpose is narrowed. How?

After speaking with his contacts at Urbana Design, Curt offers each team member the name of an architectural or engineering firm that Urbana has recommended Lowe & Company look at in their search for a partner on this project. The name Curt gives you is Techniks, located in Vancouver, Canada. Your job is to locate background information on Techniks, including its size, its ability to collaborate on a project (particularly this far from its home base), its specialties, and specifically its experience with suspension projects such as this.

Note the limitations imposed by time constraints.

You call the home office for Techniks, and the secretary, Sandy Davies, tells you that the firm's managers are out of the office for the next ten days on a job in New York City. So much for your question about distance. You explain the situation and ask if there is anyone available you might discuss the matter with. Sandy assures you that only the managers make such decisions, but if you could fax your request right away, perhaps their office could forward the fax and receive an answer by late this afternoon or tomorrow morning. You have little choice, so you agree to fax your request for information right away.

## Background Development

Consider the following questions as a way to evaluate what you've read in this case. Answer the questions on paper, in an online discussion with classmates, or work in a group and discuss possible answers orally. You may use your responses to help you in subsequent tasks.

1. Based on what you've read in the background section, how would you summarize Lowe & Company's strengths and weaknesses? Make a bulleted list of strengths on the left and weaknesses on the right. Using that list, now underline the strengths you believe will be most helpful to Lowe and Company as they try to win the bid for the city project.
2. Lowe & Company is interested in a cross-company collaboration. Make a list of 10 questions that would help you to determine whether collaboration with Techniks would be positive. These do not have to be questions that you would *ask* a representative from Techniks; rather, they should reflect what you'd want to *know* about Techniks before embarking on a short-term partnership with them.
3. In the Situation section, Maggie mentions the accident in Kansas City in 1981. In this accident, a suspension bridge collapsed under the weight of too many partygoers dancing on it. With a partner, do some brief background research on suspension bridges and accidents that have resulted from engineering mistakes. A starting place might be Henry Petroski's *To Engineer Is Human* (New York: St. Martin's Press, 1985), which has an entire section devoted to the 1981 accident. Write a brief summary of your findings for reference as you develop subsequent tasks for this case.

## Gathering More Information

Write a one-page memo to Techniks managers E. F. Black and J. St. Cloud explaining your situation and need for information. Be sure also to mark Attn: Sandy Davies.

Maggie pokes her head into your office at 2 p.m. and asks, "How are you coming?"

You explain the fact that you are waiting to hear back from Techniks and show her the fax you sent. "Well, I was hoping to have a rundown of all of these folks by tonight so I could maybe put all of the information we gather in some sort of decipherable table before tomorrow morning's meeting. If you don't hear back from these folks by 5, why don't we just scratch them off our list?"

Luckily, you receive a phone call from James St. Cloud in New York at 4 P.M. Because he is between meetings and you are facing a deadline, the conversation is brief but cordial. You are convinced, based on his genuine enthusiasm on the phone, that St. Cloud is indeed interested in the potential for teamwork on this project.

St. Cloud gives you the following information:

- His firm was established in 1964 and employs 22 full-time engineers; 20 architects (five of whom are contract employees who come on for specific projects); seven city planning/participation experts; and six management staff with professional backgrounds in engineering, architecture, and city planning.
- St. Cloud believes that the firm's specialties lie in downtown development and urban planning. He cited several successful redesign projects in cities like Stratford, Ontario; Vancouver; Park City, Utah; and Reno, Nevada.
- Techniks has successfully teamed up with a wide variety of firms on different projects including numerous public walkways and three suspension bridge projects. St. Cloud noted that several engineers employed with Techniks had a "moderate to considerable" experience with suspension techniques. In fact, he noted, one employee had been called in by the *Kansas City Star* to consult on the failure of the Hyatt Regency suspension walkway in 1981.
- If Lowe & Company is interested in meeting to discuss a team approach, Techniks could send representatives no sooner than late next week, because the firm is tightly booked at the moment with pressing jobs.

Your general sense from the phone conversation is that St. Cloud is amiable and interested in the project. However, your concerns are twofold. First, you worry that such a large firm could easily "swallow" Lowe & Company in a project such as this. While you need their expertise, Lowe & Company cannot afford to lose the high-profile potential for this job. Therefore, Lowe & Company needs to team up with an organization willing to share responsibility. You cannot discern from your brief conversation with St. Cloud whether this is possible with Techniks. Second, Techniks seems to be established or at least in particularly high demand at the moment. You worry a little about the firm's availability.

## Summary Task

Write a memo to Maggie Lowe detailing the information you gained from James St. Cloud as well as your assessment of the phone conversation. Because Maggie has told you she wants to draw this information together in some sort of readable table for the morning's meeting, pay special attention to designing the memo so that she may easily extract the particulars.

## Formal Correspondence

After a long discussion at your morning meeting, Lowe & Company employees have agreed that the best approach for the time being is to pursue the team approach. Bids for the city project are due in City Hall in three weeks. Maggie has agreed to fly James St. Cloud and an associate to Milwaukee to discuss the partnership further.

Write a letter to Techniks (attn: James St. Cloud, Manager) explaining the importance of a face-to-face meeting and the urgency of the impending deadline. Explain the offer to fly St. Cloud and an associate to Milwaukee early next week.

# Summary of Evaluative Criteria for Major Tasks in Case 17

| | *1 Unacceptable*<br>Insufficient answer to assignment expectations | *2 Below Average*<br>Inappropriate or ineffective verbal/ visual choices limit document success | *3 Meets Task Expectations*<br>Has answered objectives of assignment, but individual components could be strengthened | *4 Above Average*<br>Few flaws, document meets expectations, but could benefit from more attention to detail | *5 Excellent/ Professional*<br>Few or no flaws, demonstrates keen insight into case subtleties and details |
|---|---|---|---|---|---|
| **Purpose/Key Points**<br>▪ Identifies and defines purpose<br>▪ Articulates key points clearly<br>▪ Demonstrates audience awareness<br>▪ Summarizes key issues with attention to clarity and brevity | | | | | |
| **Context**<br>▪ Identifies/defines context and situtional contraints<br>▪ Demonstrates awareness of document situatedness | | | | | |
| **Audience**<br>▪ Identifies/defines audience<br>▪ Establishes approprite tone<br>▪ Understands technical details enough to communicate effectively | | | | | |
| **Organization**<br>▪ Demonstrates analytical insights<br>▪ Develops a clear line of argument<br>▪ Employs identifiable, appropriate pattern of organization | | | | | |
| **Design**<br>▪ Demonstrates awareness of visual design issues in correspondence | | | | | |

# PART V

# *Maintaining Professional Communication*

CASE 18

# *Trans-American Computers: International Expansion and Changing Technical Communication Needs*

You are asked to work in teams of two to create a professional oral presentation introducing the Trans-American Computers Sweetwater Operations to a group of nonnative-speaking visitors. You will also be challenged to understand the subtleties of working with nonnative speakers as well as working in a collaborative partnership.

## Background

You might compare Trans-American's beginnings with those of Gateway 2000.

Trans-American Computers, Inc., was founded in 1987 by Susan Birmingham and Lawrence Fitchburg in Salt Lake City, Utah. Not unlike several other small computer companies, Trans-American capitalized on consumers' irritation with the megacorporations that sold computer hardware and software at top dollar because of little competition. Fitchburg, a computer engineer from MIT, teamed with Birmingham, a market analyst and investor, to offer consumers no-frills computer systems at significantly lower prices than the "big names."

Note how Trans-American understood its markets and anticipated trends successfully.

Birmingham rightly assumed that by making Trans-American Computers compatible with the leading systems offered by IBM and Macintosh, the low cost would appeal to both large-order needs in businesses as well as personal computer needs at the individual consumer level. Trans-American's first print marketing campaign was simple, yet it resonated with readers. The NOW personal computer, Trans-American's first compact personal system, was featured in the ads that ran in metropolitan newspapers in Colorado, Utah, Nevada, California, New Mexico, and Arizona, the company's original target market. The half-page ads highlighted a picture of the computer monitor, on which appeared the words "The New Choice. NOW." Below the monitor the price was listed, and then the words "This is not a misprint." Some readers were amused enough with the not-so-subtle criticism of high computer prices that the ad inspired letters to the editor in Colorado and California denouncing big business monopolies. One letter went on to suggest that it was high time for a computer price war not unlike those enjoyed by travelers on major airlines.

Despite some early positive feedback on the Trans-American NOW campaign, sales did not skyrocket that first year. In fact, Trans-American operated at a loss in 1987 and recuperated only slightly in 1988. It was not until late 1988 and early 1989, after several favorable reviews of Trans-American and the NOW system appeared in *Business Week*, *Time*, and *Byte*, that sales began to make an important turnaround. In 1990, Trans-American stock jumped significantly, and the small Utah company quickly capitalized on the recent publicity to move into the black for the first time. The years 1991 and 1992 saw steady growth, and Birmingham and Fitchburg decided to expand by offering new products and expanding the geographical market to include Texas and Oregon. In 1993, two new distribution sites were added in Fresno, California and Boulder, Colorado with the main office remaining in Salt Lake City. The Salt Lake City office also remodeled by adding a second assembly and distribution warehouse and increased employment by 52 jobs.

In 1994, when several competitors were struggling with the public relations fallout from the distribution of the flawed Pentium chips, Trans-American continued to enjoy positive press, in part because the company had never purchased or distributed the Pentium chips but had opted instead for an alternative (the Omni chip) that was slightly slower but significantly less money. While Birmingham and Fitchburg secretly wrote off the "near miss" to good fortune, market analysts praised Trans-American's "foresight and strategic planning." Since 1994, Trans-American sales have continued to increase as has the company's stock value. The only brief period of decline was seen in June 1995, one month after Rudy McGonickle, Trans-American's director of sales based in Fresno, was arrested for a third drunken driving offense. When the Trans-American board of directors fired McGonickle, he publicly threatened suit. While there was only a brief mention of the incident in two magazines and newspapers, investors responded cautiously and sales dipped slightly. By July, sales and stock prices had again stabilized.

In November 1995, Trans-American decided to expand yet again and opened plants in Sweetwater, Texas and Eugene, Oregon.

## The Situation and Your Role

Your role is defined here. What does a sales associate for a computer company generally do? Where can you investigate this role further?

You are a sales associate for Trans-American based in Sweetwater. You have been with the company for two years but transferred to the new offices in Sweetwater just six weeks ago. Your immediate supervisor is Jim Shackett, director of sales. Shackett has been with Trans-American since it opened its doors in Salt Lake City, and while he is clearly committed to the company, he has made it clear that he is not happy about his transfer to Texas. Fitchburg and Birmingham sent him to Texas to make sure that the sales team for the region got off to the right start; Shackett has proven his worth time and time again. However, this move is, by all appearances, taking its toll on him, and he has frequently been short tempered and even sometimes mean spirited in staff meetings. You knew Shackett slightly when you both worked in Boulder. Since at that time you were a sales trainee and Shackett had little to do with the training program, you did not interact more than superficially; however, your respective positions did overlap peripherally for six months there. In the encounters you did have with him, you found Shackett to be an agreeable, amiable professional.

Your relationship with Shackett has been conflicted lately. Why?

Since your move to Texas, however, Shackett has publicly reprimanded you twice for relatively minor things (once it involved merely printing some information to the wrong printer, and the other time it involved reporting some sales calls incorrectly), and you have felt in general as though you and Shackett were not communicating well. You have talked about the problem with a colleague you trust, sales associate Mary Higgins, and she assured you that Shackett is simply unhappy about being in Texas. He was forced to transfer here without his family because of his wife's position as Dean of Students at a college in the Boulder area, and he is now commuting home only twice a month. This has taken an emotional toll, Higgins assured you, and Shackett's attitude about work is suffering somewhat. Higgins encouraged you to lie low and not to worry. Shackett will not be in Texas indefinitely, and his criticisms have less to do with your performance than they do with his personal situation.

On Monday of this week you walk into your office and discover a short note awaiting you on e-mail.

> *I am calling a meeting with you, Mary Higgins, and Keith Jones for 2 P.M. I realize this is short notice and that you may already have a call or another meeting scheduled. This is important, though, and I'd like you to cancel what you may already have going. We'll meet in the conference room at 2 today. Let me know if there is any reason why you can't be there.*
>
> *Jim Shackett*

Curious, you pop your head into Mary Higgins' office and discover she is not there, so you set about reorganizing your calendar for today in order to make room for the 2 P.M. meeting. Just before you leave for your first sales call at 9:30, you run into Mary in the hall.

"Did you get the e-mail from Jim?" you ask.

She nods. "Any idea what we're meeting about? His note had a tone of ..."

"Urgency?" you suggest. "I thought so, too, and I hoped you'd know what was up."

She shakes her head. "I guess we just show up with notebooks in hand and smiles on our faces."

When you arrive at the conference room, Shackett is not yet there; however, Mary Higgins and Keith Jones, a software specialist, are already seated across from one another at the table. They are reading a one-page letter, your copy of which lies on the table at the seat next to Mary's. You pick up the letter and read it.

You finish scanning the letter as Jim Shackett walks in.

"Thanks for coming, folks. I know it was probably a hassle to reschedule your day." He shuffles some papers and sits down at the head of the short conference table. "I see you've all looked at the letter from Mexico. What's your take on it?"

You think this may be a trick question, so you hold back. Mary answers promptly, "Looks as though we need to prepare for a visit."

# Ciudad Nacozari

◆ CASA CONSISTORIAL ◆

Señor Paulo Colosio y Costilla, ayuntamiento administrador
Señor Luis Zedillo Diaz, alcalde
298 Lado Bueno Calle
Ciudad Nacozari, Méjico 20006

July 15, 1998

Señor L. Fitchburg, administrador
Señora S. Birmingham, administradore
Trans-American Computers, Inc.
P.O. Box 8557-2
Salt Lake City, Utah 84101

Dear Señor Fitchburg and Señora Birmingham:

We thank you for your gracious invitation to visit your operations in the United States of America. As we communicated when we last met in 1996 January, we maintain a strong interest in locating a new Trans-American (Transamérica) operation in the city of Nacozari. So that we may better understand your organization and its structure, we wish to accept your invitation to visit Trans-American in Sweetwater, Texas.

Because several of our city administrators will be traveling abroad for some planning sessions, we respectfully request that you schedule our visit to Texas before the end of August 1997. Our group will consist of ourselves, two other city council members, and three business owners. Only one of the business owners, Señora Maria del Caballo, speaks little English. However, because she owns several office supply stores, she will make an important addition to our group, and we will translate for her when necessary.

We have enclosed a complete list of names and addresses for the individuals in our group; however, we will continue to serve as communication liaisons until travel arrangements are finalized. We look forward to seeing you once again and to viewing your newest facilities.

With best wishes for continued prosperity,

Señor Paulo Colosio y Costilla, ayuntamiento administrador
Señor Luis Zedillo Diaz, alcalde

Your purpose is defined here. What are your challenges?

Shackett smiles a little ruefully. "Well, you're right about that, but it's not the half of it. Let me fill in some details. First of all, as you can gather from the letter, Utah is looking to expand across the border. It's a risky move, since we haven't even gone further east with our operations than Sweetwater, but Utah seems to think this offers interesting possibilities given that some of what's south is sort of an untapped market. Anyway, the home office is still in the early planning stages of all of this, and as you can see, both sides are still just feeling each other out." Shackett pauses for a moment, loosens his tie, and rolls up his sleeves—a signal that this meeting is likely to be a planning session. You are immediately glad that you cleared your schedule for the rest of the afternoon. "Mary, you mentioned that it looks like we have some preparation to do. I'd add 'quickly' to that statement. This group, along with Birmingham, Fitchburg, and a few others from the home office, are coming next week."

"Next week?" Keith interrupts. "Wow. That's fast."

Shackett runs a hand through his hair and looks weary. "Yup. Apparently, this was the only time everyone could get their schedules together. So, I've called you three together as a team to help with the preparations for the visit. I can call you two," Shackett points at you and Mary, "off sales for a few days because your territories are in good order. Keith, I need your technical eye to identify some of the things we're going to need to highlight for these folks in development. This is going to be a little like show and tell. Just so you know, some of what I'm needing here has little to do with your job descriptions. I mean, I know you folks in sales aren't necessarily supposed to be working on creating documents or planning these sorts of schedules, but right now everyone is tied up, away, or being used by the home office in some capacity or another." Shackett sighs. "Sorry. I'm getting a little ahead of myself. First off, the home office assures me that the major presentations outlining timetables and strategies will be taken care of through them. But there are some other things we do need to cover in conjunction with what they're doing. First, do all of you speak Spanish?"

Mary shakes her head, "I know German, though, if we get any German-speaking Mexican officials here." Her comment garners a laugh from Shackett.

"I took some Spanish in high school," says Keith. "But hey! That was a long time ago."

"I'm in the same boat as Keith," you add.

Shackett scribbles a few notes and mutters, "Well, hell. How hard can it be to find someone here who speaks Spanish? I won't ask you to do any translating, then, but I need your help in locating somebody around here who can serve in that capacity in case we need them. Mary, can you check with personnel and see who we might have?"

"What exactly is going to need to be translated?" Mary asks as she writes in her notebook.

"Although the group from Mexico offered to try and translate for their less proficient English speakers, the home office isn't convinced that strategy would be most effective, particularly given that at least a little of what we're showing them will involve some technical language and background with the company. So the home office has asked us to come up with someone who can shadow the group while they're here to answer questions and translate when necessary. It'll make the visitors feel more at ease. So, to answer your question, probably some product lists, some history and background of the company—I think that will mostly apply to the tour of the plant—and some of the descriptions of processes we use.

"OK, now the home office has asked us to come up with a little packet of information that we can give the visitors. We can use some of the glossy promo materials on the NOW and the JET personal printer.

In many corporations facing international expansion, questions of translating documents and communicating with non-English-speaking colleagues is a major issue. Some corporations choose not to translate documents but rather to have appropriate manuals and instructions culturally designed and written by native speakers.

Many companies cultivate a collaborative approach to problem solving at a variety of levels within the organization. The Trans-American team's brainstorming here is not an unusual practice.

Do you have any suggestions for other stuff to include in this packet?"

Keith spoke up. "I have a brochure on the new graphics package. It's called White Lightning, and it's pretty slick. I could probably speed that production up. It's due next Friday, I think. I bet I could talk to some folks and have it here a couple days earlier."

Shackett nods and writes in his notebook.

"It might be a good idea to have a map of the plant in there, so they can walk around freely if they want to," you add.

Shackett nods. "Not sure I want them wandering around by themselves, but the map is a good idea. It'll give them the layout of the complex and help them visualize possibilities for their own operation."

Mary suggests, "It wouldn't be a bad idea to give them some brochures and restaurant guides for Sweetwater. They won't spend all of there time here, will they?"

"OK," says Shackett, writing. "I think it would be smart to include a manual or two in the materials to give them a sense for the machines and the accompanying user texts. That should about do it for the visitors' packets. Now, the other thing we need to think about is preparing a little sample of what we do here—the show-and-tell part," Shackett says.

Your task is defined here.

"You and Mary seem like logical choices, not just because your territories are in good order, but because you know how to talk about Trans-American products. So I would like you two to work up a short presentation on the Sweetwater Trans-American operation, with handouts or overheads, something they can look at anyway. I'm thinking handouts would be best because they can take them with them, but you decide."

The four of you brainstorm for a while longer and discuss various duties, including arranging meetings, transportation to and from the hotel, and a check of audiovisual equipment in the conference rooms. Shackett takes on the responsibility of informing employees and running a check of plant facilities. You suspect that Shackett's commitment to making the visit a success has less to do with the visitors from Mexico than it does the fact that Birmingham and Fitchburg will be here from the home office. A successful "show" may help Shackett move back to Boulder faster.

After the brainstorming session, Keith and Shackett go to Keith's office to discuss various development issues and to look at the brochure he has suggested. Shackett has asked Keith to be in charge of getting the visitors' packet together.

You and Mary then discuss the presentation you will make together. You agree to create several overheads (which will double as handouts after the presentation), including a map of the facilities and a short glossary of some terms the visitors may hear throughout their visit. Mary will provide other information taken directly from her large group sales presentations.

## Background Development

Consider the following questions as a way to fine-tune your understanding of the case and its details. Answer the questions alone or work in small groups of two or three to discuss the answers. Feel free to draw on your responses to inform any of the tasks that follow.

- Many U.S. companies are looking to expand across American borders as the most efficient means of opening new markets. As a result, many companies like Trans-American Computers are faced with new demands to translate technical material into other languages or to create new materials like manuals and instructions in several different languages at once. Find a set of instructions (e.g., computer installation, exit instructions for a commercial airline, or safety instructions for appliances) and look at how many different languages are offered by the document. How do the instructions look similar and different (aside from the obvious character and word differences)?
- Technically, you are a sales associate for Trans-American. Normally, your job would not entail the sorts of tasks you are being asked to do to prepare for the visit from the Mexican contingent. However, there are times when stretching beyond your job description is a wise move. Why is it a wise move in this situation? How would you define your purpose in this case scenario?
- You have at least two audiences for the materials you are developing in this case. Who are they? Can you define more than two? What are their expectations for this presentation? Create a list of expectations you can anticipate for each prospective audience group.

## Gathering Information

In case appendix A you will find a list of terms appropriate to the kinds of products that Trans-American markets. There are clearly too many terms here for the visitors to wade through and completely understand. You must choose an appropriate number of terms that will serve as a shorthand glossary for the nonnative speakers. Write a short e-mail message to Mary explaining which terms you've chosen, why you have chosen them, and how you plan to organize them.

## Organizing and Designing Information

Using the terms you have chosen, look through a number of computer manuals from different companies and compose a brief description (no more than 10 words) of each term, making sure that your definition is as applicable to a variety of computers as possible (remember that Trans-American is compatible with both IBM and Macintosh computers). After you have defined each, organize and design the pages so that they are consistent with one another and easy to access.

## Major Task 1

Because you would genuinely like to impress Shackett (not to mention Birmingham and Fitchburg), create a one-page overhead slide that lists some of the terms you have chosen and their corresponding Spanish term. You may also include the English definition of the term, but do not feel compelled (unless you really want to) to provide the full Spanish definition of the term. If you do not speak Spanish, there are any number of ways you may find the corresponding Spanish term. There are numerous Spanish dictionaries that may be of some help. You may also find a native speaker to help you with this task. Or you may find corresponding Spanish versions of computer manuals. Keep in mind that this is a single overhead slide and will be used in the oral presentation you and Mary are making together. Design the slide with this presentation in mind.

## Major Task 2

In case appendix B, you will find a bare bones map of the Sweetwater plant, labeled appropriately. You don't find the map particularly appealing and you also want to make the map into an overhead with Spanish terms replacing English. Experiment with redesigning the map for an overhead slide using Spanish terms to identify areas. Use the sources you previously located to help with the Spanish language.

## Major Task 3

You and Mary have divided up responsibilities for the oral presentation you will be giving your Mexican visitors. You have agreed to provide the introduction to Trans-American Computers, including its history and development over the years and an outline of its Sweetwater operations. Using the information provided in the Background section of this case, the map included in the case appendix B, and any other information about the area you may uncover through research, develop a 15-minute presentation about Trans-American and the area that will interest your audience of Mexican visitors. Keep in mind the purpose of their visit and feel free to elaborate on details as it seems appropriate. Mary will be providing the part of the presentation that focuses on the actual Trans-American product, so your focus should be on the nature of the company and its location. Feel free to develop visuals as they are appropriate to the material you choose to cover, and make sure your presentation does not go over 15 minutes. Remember that although you are certainly aiming to inform your visitors about Trans-American operations, you are also working to persuade them of its viability and potential in Mexico.

## Follow-up Task

The visit was a success, and three days later Shackett comes to your office and asks you to compose a thank-you letter for the group's visit and interest in Trans-American. Shackett plans to write one as well, but he would like your letter to reflect the rest of the team's thanks. Both Mary and Keith are out traveling this week, and it's important to get the letter out in a timely fashion. Using the address included on Costilla and Diaz's original letter, create a short letter thanking the contingent for their attention to Trans-American and for making the visit enjoyable for everyone.

# Case Appendix A: List of Possible Glossary Terms

active device
application package
area
assembly
baud rate
buffer
byte
cable
chooser
click on/off
control
design
desk accessories
desktop publishing
device
distribution
dpi (dots per inch)
drive (disk drive)
features
file
finder
folder
font
graphics
hard drive
hardware
icon
installation
interface cable
internal (drive)
keyboard
keypad
local area network (LAN)
maintenance
manual
market
memory
menu
modem
monitor
mouse
network
node
package
page (setup)
parity
pitch (or cpi)
port
printer
print resolution
quality
repair center
resolution
screen
select
serial port
software
stabilizer
startup
storage
system assembly
system file
technician
version
warranty
window

# Case Appendix B: Map of the Sweetwater Plant

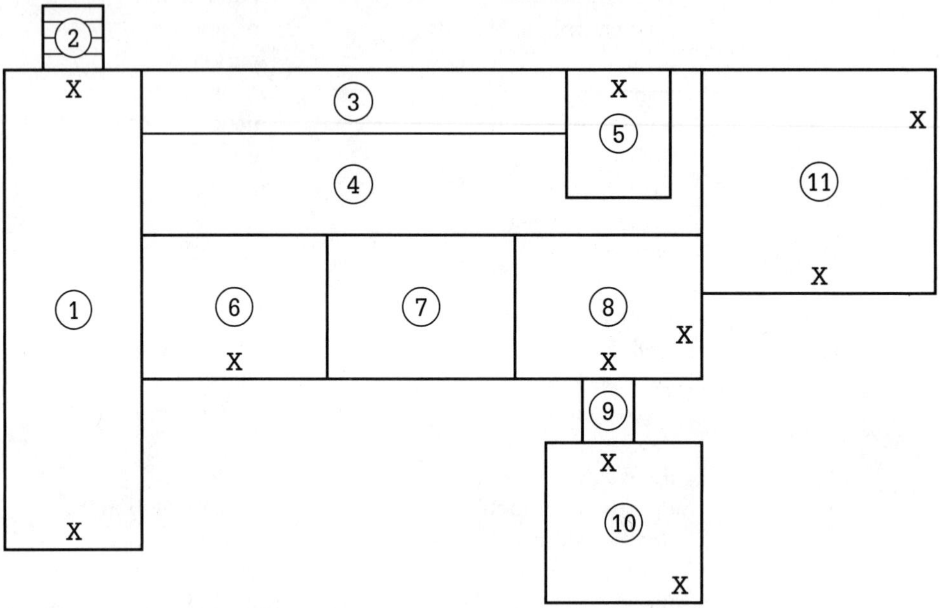


**Key**

1. Storage and loading
2. Loading dock
3. Keyboard assembly
4. monitor assembly
5. Electronics
6. Printer
7. Printer test
8. Software test
9. Walkway
10. Offices/Payroll
11. Storage

X = exits

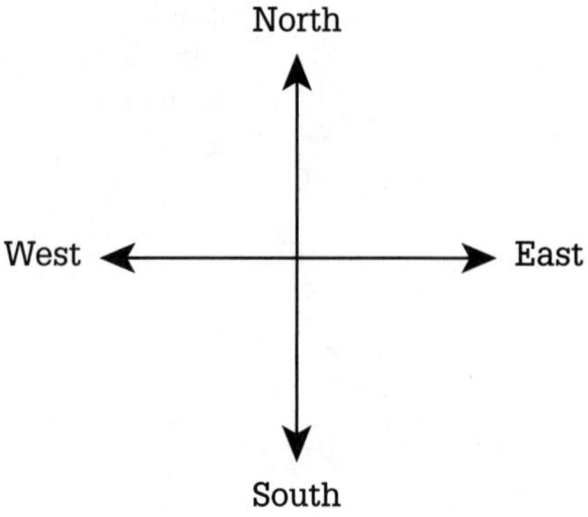

# Summary of Evaluative Criteria for Major Tasks in Case 18

| | *1 Unacceptable*<br>Insufficient answer to assignment expectations | *2 Below Average*<br>Inappropriate or ineffective verbal/ visual choices limit document success | *3 Meets Task Expectations*<br>Has answered objectives of assign-ment, but individual components could be strengthened | *4 Above Average*<br>Few flaws, document meets expectations, but could benefit from more attention to detail | *5 Excellent/ Professional*<br>Few or no flaws, demonstrates keen insight into case subtleties and details |
|---|---|---|---|---|---|
| **Purpose/Key Points**<br>▪ Identifies and meets purpose<br>▪ Articulates key points clearly<br>▪ Demonstrates research skills and employs technical languge appropri-ately | | | | | |
| **Context**<br>▪ Identifies/defines context and situa-tional constraints<br>▪ Demonstrates sen-sitivity to non-native speakers' communication needs and situation | | | | | |
| **Audience**<br>▪ Identifies/defines audience and meets nonnative speaker needs<br>▪ Establishes appro-priate tone<br>▪ Researches techni-cal terms and illus-trates them clearly for ESL readers | | | | | |
| **Design**<br>▪ Demonstrates awareness of visual design elements of task<br>▪ Overhead display(s) demonstrate atten-tion to detail and readability<br>▪ Designs are accurate<br>▪ Demonstrates an awareness of design options & technological aids in the development of these options | | | | | |

CASE 19

# *Communicating STORM-FEST Data: Reading and Communicating Technical Information*[1]

This case challenges you as a technical writer to carefully assess and consciously analyze the steps you need to use in order to understand technical, scientific data. You are asked to read several days of daily log from an atmospheric research project and rewrite the logs for a different audience.

## Background

Consider the implications for the kind of research that UCAR-OFPS does. Can you think of any situation in which *you* might directly benefit from this kind of research?

The Office of Field Project Support (OFPS) for the University Corporation for Atmospheric Research (UCAR) serves as an interagency resource for coordinating, planning, and implementing domestic and international field projects about weather patterns. Since 1984, UCAR-OFPS staff (numbering approximately 300 in the central site in Boulder, Colorado, and more than 500 elsewhere across the country and internationally) have participated in scientific planning, field deployment, operations and/or data management for a wide variety of international and national field research programs. Among the most notable projects UCAR-OFPS has completed are "Verifications of the Origins of Rotation in Tornadoes Experiment" in the southern Plains in 1994–1995; the "Global Energy and Water Cycle Experiment" in the Mississippi River Basin in 1993; and the "Central Equatorial Pacific Experiment" in the central Pacific in 1993. The data gathered by these and other experiments provide a wide variety of meteorological bases, university researchers, science foundations, and agencies such as NASA and the Federal Aviation Association with valuable information upon which to base future technical strategies or decisions. For example, a UCAR-OFPS experiment may yield crucial information about consistent wind patterns in an area proposed as the site for a new airport. If wind patterns indicate consistently unfavorable landing conditions through the majority of the fall and

[1] The author wishes to thank the U.S. Weather Research Program and the Office of Field Project Support for the University Corporation for Atmospheric Research for important contributions to the background, appendix, and website information in this case, which was derived from the STORM-FEST operations manual. The situation and characters are hypothetical.

winter months, such a report from UCAR-OFPS might have an important impact on whether the location in question is the best choice for the airport.

Because UCAR-OFPS works in conjunction with many organizations, both domestic and international, the research organization must cultivate and maintain efficient communication using a variety of traditional means (e.g., postal service and telephone) as well as advanced telecommunication and electronic means (e.g., teleconferences, electronic mail). Therefore, project managers also serve as technical communication specialists in that the data collected in a given project must be designed for a variety of audiences and purposes.

Note the defined purpose of the STORM-FEST project.

The U.S. Weather Research Program, for which UCAR-OFPS has coordinated several large-scale field projects, recently proposed a short-term project to study the structure and dynamics of winter storm fronts and associated precipitation over the central United States. There may be the promise of a comparative Central Europe study, depending upon the findings of this project. As a project coordinator at the U.S. Weather Research Program explained, this project, called STORM–Fronts Experiment Systems Test (FEST) is designed to "investigate the structure and evolution of weather front patterns and associated precipitation, as well as to test various new observational systems and operational procedures—at places like the National Severe Storms Forecast Center in Kansas City, Missouri, for example—which may be used in future experiments."

Examine the bulleted list, note any of the technical terms that may be unfamiliar to you, and look them up.

Therefore, the field phase of STORM-FEST is composed of three related components:

- An investigation of the structures and evolutions of fronts and associated mesoscale phenomena (e.g., wind circulation and cloud patterns that are 1–100 km in horizontal extent), with emphasis on precipitation and severe weather
- A research assessment of new operational and research meteorological instrumentation, facilities, composite observational networks, and other storm elements
- A study to begin to understand mesoscale prediction capabilities and limitations in active frontal regions

Because of these three primary components of the STORM-FEST project, UCAR-OFPS must manage a variety of data collection methods and sources. Using over 60 major research sites for data collection, the STORM-FEST operation relies on the following observational sources:

- Surface data collection
- National networks data collection
- State and regional networks data collection
- Research networks
- Composite surface data sets—data collected and compiled from the national, regional, and special research network data sets

- Upper air data—provided by conventional systems (National Weather Service, the military, and Atmospheric Environment Service) and research
- Surface-based rawinsonde (weather balloon upper atmosphere radar tracking)
- Aircraft-based dropwinsonde (radiowave upper atmosphere radar tracking parachuted from aircraft)
- Profilers
- Composite upper air data set—data collected and compiled from the national, regional, and special research network data sets
- Radar data—obtained from national, regional, and research networks
- Satellite data
- Aircraft data
- Model data

As you might assume, managing huge amounts of data from well over 60 sites is a daunting task for any one organization. However, such weather information (and an assessment of the effectiveness of reporting methods) is crucial to many components of everyday life such as transportation, health and medical services, even the economy. Therefore, UCAR-OFPS and the U.S. Weather Research Program Office have several teams that focus specifically on compiling all data and summarizing it for a variety of purposes. While NASA and the FAA may require the STORM-FEST data for one purpose, the National Weather Association may need to report it for a different purpose.

## The Situation and Your Role

Your role in the case is defined here. What do you see yourself doing in this role? Discuss with classmates.

You are a project assistant/technical writer for the severe weather forecasting STORM-FEST team at UCAR-OFPS in Boulder, Colorado. Your job is to help collect and sort the atmospheric and surface data from the various networks into the STORM-FEST data management system, the Cooperative Distributed Interactive Atmospheric Catalog System (CODIAC). CODIAC's major components include a data set guide, project information, station information, order entry/data delivery, inventory information, data set notes, reference contacts, and daily weather maps.

Consider the benefits of broadening accessibility of technical information to a more general public. How does the public gain?

Your direct supervisor is Sam Seidman, who has worked with the U.S. Weather Research Program for 12 years and is an experienced meteorologist. Sam has made it clear that he believes it is important to make UCAR-OFPS and U.S. Weather Research Program data accessible to a wide variety of audiences. Sam's argument is that while major financial support comes from stalwart sources such as the National Science Foundation, most of the foundation grants and federal monies have been tapped to their limit. Sam believes that to maintain and enhance experiments such as STORM-FEST, UCAR-OFPS and the U.S. Weather Research Program must not only retain the federal monies they annually receive but also pursue private foundations and individuals for new grants sources. So you are not surprised when Sam approaches you one afternoon with an idea.

"I've been thinking about CODIAC and the people who ultimately read our data over the Internet. Who do you suppose picks up on it?" he asks.

You think for a moment. "Well, it's accessible to anyone with access to the Internet, but it's designed mostly for researchers and meteorologists. I'm sure there are some serious weather buffs who check in sometimes, too, though."

The audience for the Internet information is defined here.

Sam waves his hand as though swatting away an insect. "It was sort of a rhetorical, question, actually. We both *know* that it's mostly specialists this data is put out there for. But you know, I've been interested in seeing who else we can actually get excited about our projects. And I started thinking a little about the daily weather log information we've been collecting for STORM-FEST. Right now we're not even putting the summaries on CODIAC because the information can be found throughout the rest of the data sets."

You begin to see where Sam is going and you smile. "But the weather summaries are just the types of logs you think might speak to laypeople about what we do?"

Here the purpose is narrowed.

Sam nods, excited. "Yeah, and not just your average Joe out there. I mean the audience for these will probably still be scientists and researchers to some degree, but I think maybe there's a way to offer these logs in a few different ways to different audiences. On CODIAC, people could pick the level they want to come in on. But this doesn't have to be limited to CODIAC. We could also use this in print material and certainly in the periodic reports. I think if we have a couple of different expertise levels, we're naturally going to expand. At the very least, more people may have a sense for what we do."

"It's a little like writing a manual in a couple of different languages for a multinational corporation," you reply.

"Exactly—only I think some of the differences in the way you present the data are going to be more subtle," Sam says.

You knew this was coming. "The way *I* present the data?"

Sam nods. "You're the logical choice, especially since you come out of technical communication. Why don't we do it this way? Why don't you take a week to 10 day's worth of logs and see how many ways you can write the setup? When I get back from Oklahoma next week, we'll sit down to talk about what you came up with. I think it's important you aim at least one version for an educated layperson audience—you know, readers who may not be meteorologists or researchers, but who certainly can read at the college level. Beyond that, you may want to try a couple of different audiences, like businesspeople or a college class. Use your best judgment and keep in mind that we're also thinking about future donors."

The two of you chat a while longer, and you agree to write up two versions, noting that you can't really afford to take time away from your other duties. You decide to take five consecutive days from STORM-FEST's first 10 days of summaries as your starting place. Sam assures you that you can do what you want with them as long as the information isn't changed. You may experiment with visual design as well as work with the "story" the logs are trying to illustrate. You may also construct maps if they are helpful. You have one week to produce your drafts.

It is important to find a balance between trying to understand the technical terms and not allowing your lack of meteorological experience to interfere with completing the tasks.

In the following appendices, you will find the 10 daily weather summaries as well as a glossary of relevant acronyms. You will need to look up technical terms, and there may also be abbreviations that may seem confusing to you (e.g., "prog" for program). Use your best judgment on these abbreviations and move forward if you cannot figure them out. Remember, you have only a week to have a viable draft ready for Sam, and you have plenty of other things at work to worry about. This is not to suggest that your work should be sloppy. On the contrary, you have a great opportunity here. It is important to keep in mind, though, that while it is important to use all available resources to answer a question about a term or abbreviation, if you cannot find your answer, don't let yourself be held up by it.

## Background Development

Consider the following questions as a way to evaluate what you've read in the case. Answer the questions on paper or on line in a discussion group, or work in a group and discuss possible answers orally. You may use your responses to help you in subsequent tasks.

- Consider your purpose(s) for this case. Outline briefly your primary and secondary goals. What do you hope to accomplish by offering technical information to laypeople?
- In a small group, discuss the effects of technical language on the readability of the logs. Considering the audience targeted in the case to read the logs, do you believe the technical language is an asset or detriment to your purpose? Can it be both?
- Based on what you read here, what kind of relationship would you say you have with your boss? What can you assume about Sam? How is that relationship important to your process?
- The Internet provides numerous opportunities for additional research and comparisons. Access a weather news group or meteorological World Wide Web site and examine the information provided. Briefly analyze the tone and technical level. Who would you say is the intended audience for the site? Is there anything in this site that can help you develop your own materials?

## Gathering More Information 1

The World Wide Web site **http://www_STORM-FEST.com** offers a number of maps and other background materials designed to help you get a more complete picture of your project and the technical information associated with STORM-FEST. Access the site and make some notes for your own use about the nature of the STORM-FEST project and the types of information the project is attempting to uncover.

## Gathering More Information 2

Read and carefully assess what the 10-day log for STORM-FEST reveals. Underline any words you don't immediately recognize or understand. The underlined terms should help you to be more organized when you look up definitions. Circle abbreviations that aren't immediately recognizable, so that you can more easily go back and reread a passage once you've filled in the contextual "blanks." Then, for your own reference, briefly summarize in a sentence or two what you think each of the daily logs is saying—what happened that day? This "shorthand" reading should help you to remember key points from the text.

## Major Task 1

Write two alternative five-day logs that communicate the same information, but do so to different audiences. You may choose any consecutive five-day period from case appendix A. Case appendix B provides a list of common acronyms and their corresponding technical terms that you will read in the logs.

The first alternate audience should be the audience Sam Seidman has asked you to pay most attention to. You may assume that these readers have at least a college education but are not professional scientists or researchers. These are educated layreaders who will likely have an interest in STORM-FEST, but each person's interest will vary according to his or her background and motivation for reading the summaries. The second audience may be one you choose, keeping in mind only that whomever you choose will at least have the potential for interest in STORM-FEST. You may make this audience as specialized, technical, or nontechnical as you please, but you must define these readers carefully in a cover memo to Sam.

## Major Task 2

Sam Seidman is pleased with the "lay-logs" you've created. He has now encouraged you to write a proposal to a private foundation called ADVANCE. The foundation is interested in funding projects specifically designed to create "inclusive" or "broad-based" educational endeavors in the fields of science and mathematics. The request for proposals is located in case appendix C. Respond to the call for proposals by requesting funding for a half-time person ($16,000 renewable annually) whose job it would be to translate available meteorological data to nonexpert audiences, including (but not limited to) the "lay-logs" you have created. In answering this task, you must offer other ideas for reaching lay audiences with your information in addition to acknowledging the *need* for reaching nonexperts.

# Case Appendix A: Weather Summaries February 1–10

**1 February**
On this first day of STORM-FEST, the long wave ridge to the west of the STORM-FEST experimental area and the trough to the east dominated the weather, as it had over the past month. There was a weak stationary front extending from eastern Kansas through central Nebraska extending into the Dakotas. No precipitation or active weather was expected along the front. A cold front located in western Montana was expected to move into the STORM-FEST domain on 2 February. Again, little or no precipitation was expected with the front.

The forecast progs indicated that by 0000 UTC, 3 February, the 90% relative humidity contour should move into northeast Oklahoma. This, along with a weak cold front and a weak upper-level disturbance, could trigger some stable precipitation over Texas and Oklahoma with the possibility of a few embedded thunderstorms. The amount of precipitation that falls north and northeast of Oklahoma depended on the speed of the front and the availability of moisture.

**2 February**
The general weather pattern from yesterday, 1 February, continued over the STORM-FEST domain. The stationary front located over the central United States had dissipated. There was a weak trough extending north-south from the central Dakotas down through central Nebraska and western Oklahoma. A weak low-pressure center was located over southern Canada, with a cold front extending down through Montana and Wyoming. This weak front was expected to move to the southeast over the next 24-h and provide the first frontal passage and possible precipitation in the STORM-FEST domain.

**3 February**
The weak low-pressure center that was located over southern Canada yesterday, 2 February, moved to the southwest and was located over southern Minnesota at 1200 UTC. A weak cold front extended southwestward from the low and at 1200 UTC was located over eastern Nebraska, central Kansas, and western Oklahoma. This cold front was expected to continue to move to the southeast and pass through the STORM-FEST boundary layer domain by early afternoon (2000 UTC). Little or no precipitation was expected along the front. The forecast progs indicated that this surface low and associated cold front would continue to move to the east over the next 24-h with possible precipitation in Texas and Oklahoma.

The NMC models agreed in keeping significant precipitation out of Oklahoma, instead focusing heavy precipitation over the west Gulf Coast. However, the MM4 model forecasted approximately 3 cm of precipitation near Wichita Falls, Texas, over the first 24-h forecast period from 1200 UTC. MM4 also forecasted a cm or so of precipitation in western Kansas, near where precipitation was occurring this morning, and maintained a stronger frontal structure than other models. It forecasted much colder air moving southward across Iowa, behind the shortwave trough, whereas the NGM model moved the cold air southeastward across the Great Lakes.

**4 February**
A strong surface high pressure area moved into the STORM-FEST domain, with a 1032 mb center located over Wyoming and Montana. The surface low-pressure area and associated cold front that was the initial focus of IOP 1 continued to move eastward out of the STORM-FEST domain. Split flow in the upper levels (with the northern branch in southern Canada and the southern branch in old Mexico) left the STORM-FEST domain with generally fair weather conditions.

At the surface, a second cold surge had moved south through Kansas and Oklahoma but did not produce any precipitation. Significant moisture was confined to south Texas, with an old polar continental front cutting off the return of Gulf moisture. Cool and generally clear conditions prevailed through the day over most of the STORM-FEST domain with some in the west and southwest part of the STORM-FEST domain.

A small-scale cyclonic vortex over Colorado could present a weak but interesting meteorological situation for Colorado during the next 24-h. Weak stability was allowing some amplification of this system, but this was not expected to last beyond 24-h as this system moved eastward into Oklahoma. Clear skies forecasted for northeast Kansas should allow for a radiation research aircraft flight planned for tomorrow, 5 February.

The forecast progs indicated that fair weather was expected to continue for the next several days over the STORM-FEST domain. Another trough should enter the STORM-FEST domain from the north during the next 24- to 48-h. Upper-level dynamics were expected to be strongest over the western Great Lakes, with frontal characteristics rather diffuse over Nebraska and Kansas. This continental polar airmass was relatively dry and precipitation was not expected.

**5 February**

The strong high-pressure area that dominated the weather over the past several days continued to move to the east. Once again, as on 2 February, a weak surface low-pressure area was located over southern Canada, with a cold front extending into Montana and Idaho. This surface low and associated cold front was expected to move to the southeast along the back side of the upper level trough, and be in the STORM-FEST domain tomorrow, 6 February.

**6 February**

The surface low-pressure center that was seen on 5 February over southern Canada continued to move to the southeast and was located over Lake Superior at 1200 UTC. A cold front extended southwestward from the low, through Wisconsin, Iowa, Missouri, and Kansas. This cold front was producing some very light rainfall over the central and eastern regions of the STORM-FEST domain.

Arctic air was located much further to the north of the front over the Canadian prairies. A building surface ridge just east of the Canadian Rockies was expected to push this Arctic air into the eastern and central portions of the STORM-FEST domain in the next 24- to 36-h period. This should be a strong baroclinic feature, but no significant precipitation was expected.

**7 February**

A large high-pressure area continued to dominate the STORM-FEST region. A secondary cold front was beginning to move through Nebraska and Iowa at 1200 UTC and was expected to move through the region early on 8 February. There was some possibility that there might be some very light snow over the front range of the Rockies, but no other precipitation was expected in the STORM-FEST domain.

The next 24- to 48-h was forecasted to be rather quiet as lee troughing began to push the frontal boundary into the plains. Late in the period, several short-waves in the upper-level flow could begin to interact with the boundary in the eastern half of the STORM-FEST area. Low-level moisture along and east of the warm front was expected to increase as the northern wave approached and warm advection set up. At this time, a light snow event was forecasted in the next 48- to 72-h over the Iowa/Missouri border and into central southern Illinois with this system. Some snow might begin as early as Sunday morning, 9 February.

**8 February**

Another strong high-pressure area centered over southern Canada was pushing cold air over most of the STORM-FEST area, with a stationary front backing up to the front range of the Rockies. Some very light precipitation occurred over eastern Nebraska and the Dakotas.

By 1200 UTC tomorrow (9 February), precipitation was expected to break out along and ahead of the front in eastern Kansas, Nebraska, Iowa, and western Missouri although precipitation amounts were not expected to be significant. The MM-4 model indicated increasing cloud moisture in the eastern STORM-FEST area, suggesting icing could be a problem over eastern Kansas, eastern Nebraska, northwestern Missouri and southwestern Iowa for the 1200 UTC- to- 1800 UTC, 9 February period.

By 1800 UTC, 9 February, the front was forecasted to be across eastern Kansas, Nebraska, and Oklahoma. Low-level warm advection should be increasing through the central plains with good isentropic lift over the frontal boundary. Early precipitation along and east of the front would be mainly light, but could be enhanced by an approaching Pacific shortwave. It did not appear that there would be any jet streak influence with this event, since there appeared to be no indication of a speed max associated with the shortwave.

**9 February**

The strong surface high pressure area seen yesterday, 8 February, had moved to the southeast and was centered over Illinois. A weak stationary front was located over western Kansas, Nebraska, and Wyoming and was expected to move slowly eastward to extend from northeast Iowa to eastern Arkansas by 0000 UTC tomorrow, 10 February. At 1200 UTC, light precipitation was occurring over eastern Nebraska and Iowa.

This front was forecast to move eastward and lie north-south through eastern Illinois by 1800 UTC, 10 February. Broken low-level clouds with some isolated showers were expected to occur near the front throughout the period.

A new surge of Arctic air was currently poised over northern Montana and was expected to push into the extreme northern STORM-FEST area by 0000 UTC, 10 February. The front was forecast to move southeastward and be in the Oklahoma/Arkansas area in 48-h. The depth of the Arctic air should be limited to below 700 mb but should deepen dramatically over the extreme northern STORM-FEST region. The airmass should be quite dry with only a few showers expected near the front, especially over the eastern STORM-FEST areas.

There was some concern that the long-range models may not be initializing the systems off the west coast very well. Generally, the models were forecasting a push of more energy into the southern stream, south of the STORM-FEST area. This would keep the cold front from moving much further southwestward. The magnitude of this impulse was uncertain, thus leading to uncertainties in the frontal position in the next 24 to 48-h.

**10 February**

The weak Arctic cold front that was discussed yesterday, had moved down through the northern portion of the STORM-FEST domain from a low-pressure center located north of Lake Superior. The front was forecasted to weaken during the afternoon. Precipitation had developed and moved into Illinois, and a small area of rain and freezing rain developed well southeast of the front in northeast Arkansas, in an area of isentropic uplift (seen in the 296K surface) and a nicely collocated PV anomaly at 320K. Both areas of precipitation were forecasted to weaken and propagate eastward through the period. No precipitation occurred in the CP-3 and CP-4 dual-Doppler area.

By 0000 UTC tomorrow (11 February) the front was forecast to extend from southeastern Wyoming to near the STORM-FEST boundary layer network. There might be some freezing drizzle generated in the low-level clouds behind the front. Precipitation accumulation was expected to be insignificant in the next 24-h.

A major shortwave has moved into central and south California, driven by a 120 kt jet, and should begin to affect the STORM-FEST domain by 0000 UTC tomorrow (11 February). AVN grids indicate two areas for greater than .01 inch of precipitation: one in Texas along the Gulf Coast and the other centered on Arkansas, Missouri, and Oklahoma, with a .03 inch max 6-h accumulation ending at 1200 UTC.

# Case Appendix B: Acronym List

| Acronym | |
|---|---|
| IOP | Intensive Operation Period |
| MM4 | NCAR Mesoscale Model (version 4) |
| NGM | Nested Grid Model |
| NMC | National Meteorological Center |
| NWS | National Weather Service |
| UTC | Universal Time Coordinated |

## Case Appendix C: ADVANCE Request for Proposals

**BACKGROUND**

ADVANCE, a private foundation dedicated to the advancement of technical and scientific education, is particularly interested in projects that seek to broaden the general public's interest in the sciences. ADVANCE recognizes that science and mathematics as disciplines can be exclusive. Historically, for example, women have been discouraged from entering certain technical fields. ADVANCE aims to support projects that target previously excluded or under-represented groups. Below is the required format for successful proposals with ADVANCE. Awards from $500 to $20,000 are made quarterly in January, April, July, and October. ADVANCE reserves the right to award part or all of the requested funding for a given project. Proposals longer than 2, 500 words will not be considered.

**REQUIRED FORMAT**

- **Brief introduction** including background of organization, individual, or group seeking funding. ADVANCE is interested in the level of commitment the proposer has to education.
- **Outline of proposed project** as well as an argument for its necessity.
  1) Purpose and goals for project
  2) Proposed plan for project—how does the project meet a specific need and what are the details of its implementation?
  3) Evaluation (how will the project be measured for its effectiveness?)
  4) Timeline for project (how long will it take to complete the project or will it be ongoing?)
  5) Costs for project
- **Brief summary of goals and intended beneficiaries**—who will benefit from this project and how?

Proposals should be postmarked no later than the first day of the month in which the proposals will be reviewed (noted above). They should be double-spaced and in letter format. Please mail three (3) copies of the proposal for reviewers.

Address your letters to
Ms. Tammy L. Fladebo, Chair
ADVANCE Grants Program
P.O. Box 33073
Raleigh, North Carolina 99023

## Summary of Evaluative Criteria for Major Tasks in Case 19

| | ***1 Unacceptable*** Insufficient answer to assignment expectations | ***2 Below Average*** Inappropriate or ineffective verbal/ visual choices limit document success | ***3 Meets Task Expectations*** Has answered objectives of assignment, but individual components could be strengthened | ***4 Above Average*** Few flaws, document meets expectations, but could benefit from more attention to detail | ***5 Excellent/ Professional*** Few or no flaws, demonstrates keen insight into case subtleties and details |
|---|---|---|---|---|---|
| **Purpose/Key Points**<br>▪ Identifies and meets purpose<br>▪ Articulates key points clearly<br>▪ Demonstrates close reading ability with appropriate use of technical language skills and design elements | | | | | |
| **Context**<br>▪ Identifies/defines context and situational constraints<br>▪ Demonstrates awareness of log/proposal situatedness | | | | | |
| **Audience**<br>▪ Identifies/defines audience and meets identifiable needs<br>▪ Establishes appropriate tone<br>▪ Reads and responds to technical information effectively<br>▪ Proposal offers appropriate information | | | | | |
| **Design**<br>▪ Demonstrates awareness of visual design elements of task<br>▪ Demonstrates an awareness of design options & technological aids in the development of these options | | | | | |

CASE 20

# *Not Just Filling in the Blanks: A Professional Review at Wolfe, Inc.*

This case challenges you to carefully analyze medical and other relevant information about an employee's work-related injury and its ramifications for the employee's future. The case includes a correspondence series from an insurance company and medical doctors. You will analyze the correspondence as well as their context and assess the best way to meet the employee's needs.

## Background

Wolfe, Inc. is a mechanical engineering contracting business located in Boulder, Colorado. Since 1972, Wolfe, Inc. has contracted for mechanical installations throughout Colorado, Idaho, Utah, and Wyoming and is well-known for its single-source responsibility. Single-source responsibility is a concept many mechanical engineering contractors strive for. It means that all elements of the mechanical installation process report from the same contractor; thus, responsibility for an entire project rests solely with Wolfe, Inc.

The single-source responsibility identity Wolfe, Inc. has cultivated stems largely from the experienced labor force representing pipefitters, millwrights, boilermakers, plumbers, sheet metal mechanics, service technicians, architectural engineers, and electrical engineers. All employees work directly for Wolfe, Inc. and contract as teams to work on various commercial, industrial, and institutional projects involving new construction, repair, or modernization of existing structures. Workers at Wolfe, Inc. are not unionized. Some of the projects Wolfe has completed in the last five years include:

Examine the different specialities listed here. How many of them are you familiar with? How many are unfamiliar? Explore the Internet or your library to gain some insight into the specialities about which you know little.

The list of projects is relatively diverse, both in scope and in area. What does this tell you about Wolfe, Inc?

- Tudor Ave. Shopping Center construction in Denver, Colorado
- Municipal wastewater treatment plant construction in Boise, Idaho; Cain, Wyoming; Thurgood, Wyoming; Park City, Utah; and Eagle's Nest, Colorado
- Bridge Science Center construction at Colorado Technical Institute in Boulder, Colorado

- Hospital addition modernization and construction at Vista, Colorado and Boise, Idaho
- A variety of office building construction, modernization and repair projects in Colorado, Idaho, Wyoming, and Utah

Wolfe, Inc. is affiliated with several professional associations including DBIA (Design Build Institute of America), AGC (Associated General Contractors of America), and ASME (American Society of Mechanical Engineers) to name only a few. Wolfe's pipefitting team was cited in 1995 by the national union for "demonstrated excellence in hiring and continuing education programs for pipefitter professionals." Indeed, Wolfe, Inc. is regarded by many in the region as one of the most reputable and thorough contractors for plumbing and pipefitting. Because of the company's dedication to its work force, demonstrated through excellent salaries and benefits packages, continuing education programs, competitive retirement package, and wellness programs, Wolfe's employees have never unionized, despite efforts to change that status on several occasions.

Jack Adams and his position at Wolfe, Inc. are introduced here.

Wolfe, Inc. employs many professional engineers and technical experts in a variety of mechanical capacities. This case focuses on Jack Adams, a pipefitter for Wolfe; thus, it is important to understand the demands of his job. Wolfe's plumbers work from blueprints or drawings that show the planned location of pipes, plumbing fixtures, and appliances. Plumbers and pipefitters work together on location and in the planning stages to engineer a specific strategy for fitting the piping into the structure of the project within the constraints of the situation (limited space and materials). They measure and mark areas where pipes will be installed and connected and anticipate possible obstructions, such as electrical wiring or existing gas lines. Depending on the needs of the project, plumbers and pipefitters may need to alter the existing structure—for example, by cutting holes in walls, ceilings, and floors to make space for necessary piping. To assemble the system, plumbers cut and bend lengths of pipe using saws, pipe cutters, and pipe-bending machines. Then, depending on the type of pipe used, lengths of pipe must be connected with fittings. If the pipe is plastic, plumbers connect the sections and fittings with plastic or rubber adhesives. Copper pipe requires that workers solder the fitting in place with heat.

Because of the physical nature of the work, plumbers and pipefitters are required to have a certain amount of physical agility and stamina, in part because they frequently must lift heavy pipes, stand for long periods of time, and sometimes work in uncomfortable, cramped, and damp locations. All Wolfe employees are subject to yearly physical examinations, periodic drug testing, and are enrolled in annual wellness clinic seminars offered by Wolfe, Inc.

Health insurance is essential for plumbers and pipefitters because of the rigors of the work and potential for accidents. Plumbing and pipefitting professionals are subject to falls from ladders, cuts from sharp tools, and burns from hot pipes or from soldering equipment. Wolfe offers comprehensive health benefits to all full-time employees, including all plumbers and pipefitters currently employed by the company, and optional family policies with a copayment.

Wolfe's health insurer covers all major medical, illness, on-the-job accidents, maternity and limited extended leave from work.

Finally, because of the contractual nature of their work, Wolfe, Inc. plumbers and pipefitters frequently spend quite a bit of time traveling to and from work sites. Depending on the location and duration of a given project, these professionals may be asked to temporarily relocate to ensure easier access to the site. This is especially true during winter months, when travel to and from sites is often hampered by poor driving conditions.

The annual employee review process is outlined here as well as the purpose behind your task.

Every Wolfe employee—upper management, project workers, and secretarial staff alike—annually meets with an employment team that includes a human resources representative, a direct supervisor, and a coworker as the first step in a performance review. This review is designed to evaluate the employee's attitudes, experiences, and performance over the past year and also serves as an opportunity to brainstorm as a team for ways to improve the work environment for all involved. The meeting is an interactive process designed to offer the employee constructive feedback on performance while also allowing for response. In addition to the performance review meeting, the human resources representative collects and summarizes performance review questionnaires completed by direct supervisors and self-analyses written by the employee.

## The Situation and Your Role

Your role and the problem are both defined here.

You are a new human resources employee with Wolfe, Inc. Your first employee review involves Jack Adams, a 39-year-old pipefitter who has been with the contractor for 14 years.

Jack's case is slightly more complicated than a normal annual review might be because his future employment at Wolfe is in question. About six months ago, Jack suffered an injury while on a job site. While working with an underground pipe crew on a municipal wastewater treatment plant project, one of the hydraulic excavators (a CAT 300L model), which was on relatively unstable and wet ground, rolled down a slight embankment, pinning Jack against the earth wall of the dig site for a few minutes. Amazingly, Jack was able to walk away from the accident with what appeared to be only a dislocated shoulder. Coworkers believed that Jack was saved by the soil that was spongy and yielding from the recent rainfall. A coworker called for an ambulance and Jack was taken to a nearby medical clinic, where he was examined and his shoulder was put back into place. The doctor warned Jack that he may also have sustained some muscle and ligament strain and that it was inadvisable that Jack return to his normal duties at work before four weeks. Jack would also require some physical therapy in order to regain complete strength in the arm.

After 20 days, Jack felt good enough to completely remove the sling on his arm and return to work. Initially, Jack was restricted from significant lifting, and instead he filled in with planning team members, marking cable locations. Five weeks after the accident, however, Jack maintained he was able to return to the regular duties of his pipefitting position. During his absence from the job,

Jack was guaranteed his wages because of worker's compensation benefits, and his health insurance covered all medical bills, sans a $250 annual deductible charge.

Jack's health complications are especially relevant to his position as a pipefitter at Wolfe, Inc. Can you explain why?

Not long after his return to regular duties, Jack began to experience some pain in his neck and numbness in his forearm. He described the initial sensation to his coworkers as "pins and needles," but later, after dropping some soldering materials, Jack admitted that he had lost all feeling in the forearm for a few moments. He attributed the problem to overexertion so soon after the accident, but continued to work. For two weeks, the tingling and numbness in his arm intermittently reoccurred. Finally, after losing his grip on a soldering tool and subsequently nearly causing an accident that might have hurt his coworker, the supervisor, Bill Raines, spoke with Jack and both agreed that the problems were perhaps not the result of simply weakened muscles and overexertion. But when Jack returned to his physician, the examination suggested that Jack's problem could be associated with some inflammation caused by overexertion. The doctor prescribed some anti-inflammatory drugs, and told Jack that if there was no improvement in one week, then he should return to the clinic for more X rays. Jack took a week of sick leave at this point with the hope that his arm would improve if he kept it more or less inactive.

Over the next week, Jack's symptoms did not subside, so he returned to the physician. The doctor scheduled X rays, blood work, and nerve conduction tests. After consulting with an orthopedist and a nerve specialist, the doctor told Jack that it was likely he had sustained some damage to his neck which in turn had damaged what looked to be the radial nerve in his arm. The doctor noted that without closer examination and possibly surgery, it would be difficult to uncover the cause of the extent of the damage. While the doctor could not be sure the problem was directly the result of the accident, the likelihood of a connection seemed strong. Surgical efforts to repair the nerve would provide a clearer indication of the extent of the damage and possibly how long the problem had existed. However, Jack could opt for continued physical and drug therapy to minimize the symptoms and manage discomfort. Doctors noted that while the nerve damage might be addressed through these options, there was also a possibility that Jack might never return to pipefitting because of the physical demands of the work. Doctors encouraged Jack to avoid physical labor throughout any physical and drug therapy combination or until after surgery.

Jack determined that his best course of action was to first undergo more physical therapy and assess his progress in a month. During this time he continued to work for Wolfe, but he did not perform his usual duties. Instead, Jack consulted on development stages of various projects, maintained records, and helped manage inventory. He frequently mentioned to supervisors and coworkers that he missed his pipefitting work on the crews and very much wanted to return to it.

Here you are introduced to some technical terms relevant to Jack's case. Identify the terms and look them up if you do not understand them.

After a month, doctors determined that Jack had made little progress and surgery was likely the best option for recovery. He agreed and underwent two procedures—the first exploratory and the second reparative—just before his annual review in April. The initial surgery indicated that Jack had sustained some damage to the brachial plexus in the axilla (under the arm). Whether the condition was connected or not, Jack had also experienced some bleeding in his arm. The blood infiltrated the nerve, and the radial nerve indicated some scarring. Doctors were unsure whether the damage in the arm had been caused by the accident with the hydraulic excavator because the scarring on the nerve looked older than four months, and it was not impossible for symptoms like numbness to surface as much as a year after initial damage occurred. The goal of the surgery was to reduce scar tissue and secure the nerve by suturing components to the muscle tissue. Surgeons determined that damage to the brachial plexus, while caused by trauma, had the potential to heal with time and physical therapy. Therefore, they chose not to do anything and reserved the option to perform a second surgery at another date.

Since the surgery, Jack has been unable to work. Wolfe, Inc. has continued to pay worker's compensation benefits, despite the fact that doctors cannot be completely sure the nerve damage was caused by trauma from the accident. Jack's doctors have indicated in a report (located in the appendix material), that it is unclear whether Jack will ever fully recover normal strength in his forearm and hand. You participated in Jack's annual review meeting in which you discussed a number of issues as well as Jack's medical status. His coworker and supervisor, also present at the meeting, were consistent in their praise of Jack's ability. Jack indicated several times that he enjoys his work and has very much missed working with the crew during his recovery. He has a strong commitment to Wolfe, Inc. and desires to return to work.

Your purpose is further defined here. What potential obstacles do you face in your evaluation of Jack's performance over the past year?

As the human resources representative on Jack's annual review committee, you are faced with gathering and summarizing the review forms, filled out by Bill Raines, Jack's direct supervisor and pipefitter foreman, and Sam Brammer, a coworker on two of the projects Jack was involved with in the past year. In your role as human resources representative, you are required to make recommendations on Jack's employment status and salary range. Normally employees receive cost of living and compensatory raises with annual reviews and may also be promoted. If an employee has been problematic (e.g., if an individual has undermined morale on a given project, endangered coworkers in any way, or supervisors have noted an unusual or unexplained amount of absenteeism or tardiness to a site), the human resources director might make a recommendation to terminate an employee or place the individual on probation. In addition, you are required to summarize briefly the face-to-face review committee meeting on the form located in case appendix C.

Your audience and further complications are introduced here.

Your report and recommendation goes to the human resources director, who then meets with Jack to discuss and finalize action. Thus, while your recommendations and observations are extremely important to the action the human resources director ultimately takes, you are not guaranteed that your director will embrace or act on everything you suggest. All final decisions about employment rest with the director, but her history is such that she shows faith in her staff and rarely ignores the recommendations  As a result, your report must summarize the others' input, explain and highlight the pertinent technical information, and persuade the director that your recommendations are fair and appropriate. Although Jack's self-analysis and the review forms are attached documents to your final report, your aim should be to summarize and highlight the thrust of their arguments as best you can for the director. Also, the medical doctor's written report, though included in Jack's personnel file, is not a part of the annual review package unless you argue for its relevance and include it yourself.

Beyond the fact that Jack's medical future is in question, Wolfe, Inc. must examine the financial benefits for keeping Jack on the payroll and hiring an additional pipefitter to complete the work Jack would normally do were he healthy. Jack currently earns approximately $46,080 annually without overtime. Legally, Wolfe, Inc. is not required to pay worker's compensation benefits indefinitely, but the company may choose to do so.

## Background Development

Consider the following questions as a way to fine-tune your understanding of the case and its details. Answer the questions alone or work in small groups of two or three to discuss the answers. Feel free to draw on your responses to inform any of the tasks that follow.

- In at least two places in this case it is important for you to understand the meaning of some technical terms. Identify the terminology in this case that is unfamiliar to you. With a partner or in a small group make a list of words that require further definition. Then divide the list and seek out the meanings of those terms. Combine your findings and use them as a glossary. Also keep track of where you locate these meanings, so that you not only have a glossary of terms, but a glossary of research sources.
- In a small group discuss the relevance of the doctor's inability to determine for certain what has caused the nerve damage in Jack's case. Should this detail have a bearing on the outcome?
- The case never offers you insight into how many workers Wolfe, Inc. employs. Is this an important detail to know for this case? Can you estimate the size of Wolfe, Inc. based on other evidence provided in the text of the case? If so, what details help you with that and why?
- With a partner or small group, diagram the annual review process steps at Wolfe, Inc. What do you see as the potential benefits and drawbacks of this approach?

## Getting Started

For your report to the director, you must summarize information offered by Jack's coworkers, offer insights into the review team meeting, and offer a recommendation about action regarding Jack's case. Clearly, this review team is faced with some extenuating circumstances, none of which the review form allows for. For your own purposes, review the background information on Jack's situation and take some notes that summarize what you see as the "extenuating circumstances" of this review. Because these are notes to help you, keep your summary informal and to only a few sentences. You may choose to share your notes orally or in class discussion with a small group or partner.

## Organizing Information

Review the documents from the case appendices. Create a list of the most important criteria Wolfe, Inc. emphasizes in the annual reviews, based on what you can discern from the documents. Alongside this list, additionally outline any criteria *you* believe ought to be valued or highlighted, but is not because of the limitations of the review forms. Use this list to help organize your ideas for Major Task 1.

## Major Task 1

Using the list of criteria for evaluation you created from the Wolfe, Inc. annual review form only, summarize the review team's comments on Jack's performance over the past year. You may use the written forms as well as the notes you took on the face-to-face review team meeting and Jack's self-analysis. You may assume that this report will go to your direct supervisor, Teri Bower, director of human resources for Wolfe, Inc., and it will serve as the primary piece on information upon which she bases her decision about Jack's employment status and salary increase. Write your report in memo form. Remember that you are relatively new to your work at Wolfe, Inc. and you know little about the tone of such reports. Though you have asked several coworkers about the tone of review summaries, most have given you little insight beyond the fact that the report goes into the personnel file, which is confidential and only open to human resources management and the employee.

## Follow-up Task

Using the doctor's report, your list of criteria not covered by the review form, and Jack's self-analysis offer an addendum to the report you constructed for Major Task 1 analyzing the extenuating circumstances of Jack's case and arguing how best to meet Wolfe's needs as well as Jack's professional needs.

## Major Task 2

After carefully examining the review form, you decide that it does not allow for enough flexibility in response. It is difficult to offer insight into extenuating circumstances like Jack's because of the limitations of the questions. Write a memo to your supervisor, Director of Human Services Teri Bower, that illustrates the form's shortcomings. Your memo should encourage Ms. Bower to allow the staff to review the form's effectiveness in order to make changes. Since you have no knowledge of the history of this form, gauge your tone and approach accordingly.

## Follow-up Task

Your supervisor has agreed to let you redesign the annual review form. In groups of three or four, serve as a human resources "focus group" and brainstorm for ideas about what to change. Collaboratively, then, draft two options for a new annual review that you could present to Ms. Bower.

## Case Appendix A: Employee Review Forms

# Wolfe, Inc.
## Employment Review

Thank you for agreeing to participate in an annual employee review. Your insights into the performance of Jack Adams are important to the overall assessment of the Wolfe, Inc. community. This review is designed to help create a constructive and well-rounded picture of the employee and his or her strengths and weaknesses. Your comments will remain confidential on this form; however, you are encouraged to share your insights openly and honestly at the review team meeting. The process is designed to encourage constructive feedback from which the entire company should benefit.

If you feel unable to provide insights on a given questions or questions, mark N/A in the appropriate space. If you believe you cannot provide any useful feedback and/or are unable to offer objective assessment, please speak with the human resources representative on the team and he or she will appoint an alternate reviewer.

### Reviewer Background

**Name** Bill Raines **Position/Employee Rank** Supervisor III

**Social Security Number** 481-00-4811

**Mo./Years with Wolfe, Inc.** 8

**Relationship with Reviewee (circle one):** Co-worker/Team Member, Supervisor, HR personnel, Other (please explain)

### Reviewee Background

**Name** Jack Adams **Position/Employee Rank** Pipefitter II

**Social Security Number** 484-82-5553

**Review #** 13

### Please Circle Your Answer to the Following Questions:

1) **How would you rate the reviewee's attention to promptness?**

Excellent Good Fair/Average Poor Unacceptable

2) **How would you rate the reviewee's concentration while on the job?**

Excellent Good Fair/Average Poor Unacceptable

3) **How would you rate the reviewee's ability to work as a productive team member?**

Excellent Good Fair/Average Poor Unacceptable

**4) How would you rate the reviewee's quality of work?**

<u>Excellent</u> Good Fair/Average Poor Unacceptable

**5) How would you rate the reviewee's ability to interact positively with co-workers?**

<u>Excellent</u> Good Fair/Average Poor Unacceptable

**6) How would you rate the reviewee's attitude about his/her job?**

<u>Excellent</u> Good Fair/Average Poor Unacceptable

**7) How would you rate the reviewee's training for this position?**

Excellent <u>Good</u> Fair/Average Poor Unacceptable

**8) How would you rate the reviewee's ability to see a project to its completion?**

Excellent Good Fair/Average Poor Unacceptable <u>N/A</u>

**9) How would you rate the reviewee's individual contributions to projects with which he/she has been involved?**

Excellent Good <u>Fair/Average</u> Poor Unacceptable

**10) How would you rate the reviewee's potential for leadership at Wolfe, Inc.?**

Excellent Good Fair/Average Poor Unacceptable <u>N/A</u>

**Briefly explain what you see as the reviewee's best assets (biggest strengths).**

Jack's co-workers like him a lot. He cares about his work and does good quality work. I know he really wants to continue his pipefitting duties, and if he can he should be able to.

**Briefly explain what you see as the reviewee's weaknesses (needs for improvement).**

Jack got hurt a few months back, so judging his ability to finish a job isn't really fair here. His health may be a problem for him.

# Wolfe, Inc.
## Employment Review

Thank you for agreeing to participate in an annual employee review. Your insights into the performance of Jack Adams are important to the overall assessment of the Wolfe, Inc. community. This review is designed to help create a constructive and well-rounded picture of the employee and his or her strengths and weaknesses. Your comments will remain confidential on this form; however, you are encouraged to share your insights openly and honestly at the review team meeting. The process is designed to encourage constructive feedback from which the entire company should benefit.

If you feel unable to provide insights on a given questions or questions, mark N/A in the appropriate space. If you believe you cannot provide any useful feedback and/or are unable to offer objective assessment, please speak with the human resources representative on the team and he or she will appoint an alternate reviewer.

**Reviewer Background**

**Name** Sam Brammer **Position/Employee Rank** P.F. III

**Social Security Number** 589-43-2478

**Mo./Years with Wolfe, Inc.** 7

**Relationship with Reviewee (circle one):** (Co-worker)/Team Member, Supervisor, HR personnel, Other (please explain)

**Reviewee Background**

**Name** Jack Adams **Position/Employee Rank** Pipefitter II

**Social Security Number** 484-82-5553

**Review #** 13

**Please Circle Your Answer to the Following Questions:**

**1) How would you rate the reviewee's attention to promptness?**

Excellent (Good) Fair/Average Poor Unacceptable

**2) How would you rate the reviewee's concentration while on the job?**

Excellent (Good) Fair/Average Poor Unacceptable

**3) How would you rate the reviewee's ability to work as a productive team member?**

Excellent (Good) Fair/Average Poor Unacceptable

**4) How would you rate the reviewee's quality of work?**

Excellent Good Fair/Average Poor Unacceptable

**5) How would you rate the reviewee's ability to interact positively with co-workers?**

Excellent Good Fair/Average Poor Unacceptable

**6) How would you rate the reviewee's attitude about his/her job?**

Excellent Good Fair/Average Poor Unacceptable

**7) How would you rate the reviewee's training for this position?**

Excellent Good Fair/Average Poor Unacceptable

**8) How would you rate the reviewee's ability to see a project to its completion?**

Excellent Good Fair/Average Poor Unacceptable

**9) How would you rate the reviewee's individual contributions to projects with which he/she has been involved?**

Excellent Good Fair/Average Poor Unacceptable

**10) How would you rate the reviewee's potential for leadership at Wolfe, Inc.?**

Excellent Good Fair/Average Poor Unacceptable

**Briefly explain what you see as the reviewee's best assets (biggest strengths).**

Jack works well with others. Knows his job.

**Briefly explain what you see as the reviewee's weaknesses (needs for improvement).**

His health may be a weakness but you can't judge that.

## Case Appendix B: Jack Adams' Self-Analysis

Dear Review Committee Members:

Thank you for your support and comments at our meeting yesterday. This letter is my self-analysis and response for the review package.

I have worked on two projects during my evaluation year. The first was the Vail Botanical Gardens addition, which was completed in September. The second was the water treatment project in Carlson. Although I started that project, I was injured on the site. I was not able to finish that project and have missed quite a lot of work since then due to my induries. I was out most recently with surgery on my arm.

Even though I didn't finish the Carlson project, I believe my work up to that point was as good as it ever has been. No one has ever found problems with my work. I am thorough and careful. Also, I completed one education block last January when I went to the APPA conference in Seattle with Matthews, Kjellstrom, Origer, and Cunningham. My wellness points are low this year because I haven't been able to continue lifting weights at the gym because of my injury.

I believe that I have been as productive and as active as I could be at Wolfe even though I've been out injured quite a bit. Even when I couldn't be out on the project after my accident with the CAT, I made sure that I was around and helped out in all different ways including marking and inventory. I say that because I want to emphasize that I care about my work at Wolfe and don't want to give it up. I think that the meeting yesterday showed that my supervisor and coworkers understand I'm a good worker and dedicated to what I do.

Thank you for your consideration.

Sincerely,

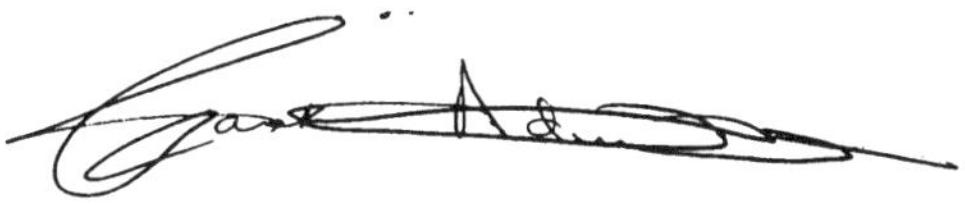

Jack Adams

# Appendix C: Notes from Jack Adams' Employment Review Meeting

| | |
|---|---|
| **Background:** | Employed continuously at Wolfe for 14 years |
| **Salary:** | $24/hour plus overtime at time and a half |
| **Wellness points to date:** | 148 |
| **Continuing Education:** | APPA Conference, Seattle (May) |
| **Status:** | Married, two dependents |
| **Description:** | Ht. 6'1"; wt. 190 lbs.; Eyes: brown; Hair: Brown |

- We began the meeting by talking about our respective positions and relationships to Jack Adams, the review. As the human resources representative I am the one person on the team with no working relationship with Mr. Adams. I acknowledged that this is unusual; however, because Ross Mitchell left our office not long ago, and Mitchell has conducted the previous four reviews with Adams, our office didn't have anyone who is especially well-versed in Adams' case.
- Bill Raines, Adams' direct supervisor, offered an overview of the projects with which they had been involved in the past 12 months. He praised Adams' work as "thorough and conscientious," and he noted that Adams has a nearly spotless record as a pipefitter with Wolfe.

  Raines suggested that Adams could perhaps use his time away from work (because of his injury) to attend some conventions as a representative for Wolfe pipefitters, and he would support such a move. Raines noted that Adams is a fine representative of Wolfe and that his coworkers respect and like Adams. When asked if he could highlight any specific weaknesses or areas in need of improvement for Adams, Raines indicated he could think of none. He noted that it was difficult to judge such an issue when Adams has been absent from work due to his injuries.
- Sam Brammer, a coworker of Adams', echoed Raines' praise of Adams' work and relationships with coworkers. He noted that Adams is well-liked and trusted by coworkers. Brammer specifically pointed to Adams' ability to pinpoint problems and offer viable solutions. Such an instance occurred in the Vail project when connection to existing structures was hampered by some electrical work that was not illustrated on the blues. Adams offered a solution that was efficient and logical. Brammer also noted, however, that he had not worked with Adams since May, because Brammer was not working with the team in Carlson. He felt he could only judge Adams' work in Vail.

  When asked if he could point to any weaknesses, Brammer did criticize Adams for taking so long to report the numbness in his forearm, noting that in one instance, Adams did risk the safety of at least on other worker when he lost control of some soldering tools. He acknowledged that he was not present at the incident, and he had only heard about it through other coworkers. Adams agreed with Brammer in this instance and I noted that it seemed the two had discussed their feelings on the incident prior to our meeting. This suggests that Adams has a frank and open interpersonal relationship with at least one of his colleagues.

- Both Brammer and Raines suggested that if Adams could not return to work because of his injuries, he would be missed. After 14 years, Adams is seen as a key team player. Adams was quick to point out that the doctors have not specifically at this juncture ruled out a complete recovery.

- We concluded the meeting by discussing the wellness program's point system, and the men disagreed with each other about the effectiveness and usefulness of the program. Both Brammer and Adams contended that the wellness merit system at Wolfe was too difficult for most employees to achieve substantial bonuses. In fact, they argued, the wellness program had become for some a bit of a joke. Raines disagreed by explaining that since the inception of the wellness merit system five years ago, absenteeism and sick leave days had decreased across the board. Adams contended that because of his injury, he would be ineligible for any wellness bonuses, and he felt this was not fair. Raines pointed out that Adams certainly did not need to lift weights in order to achieve his wellness points, but instead could consider a number of other programs including walking or a nutrition class at the hospital. Adams asked if physical therapy might count for wellness points, and Raines said he'd look into it.

- In sum, both Raines and Brammer seem to have a solid working relationship with Adams and they certainly communicate with one another easily and effectively. All three seem quite concerned about Adams' return, though I suspect both Brammer and Raines are also worried about the near accident caused by Adams' numbness.

## Case Appendix D: Doctor's Assessment

COLORADO FAMILY CLINIC & HOSPITAL

CF✚CH

To: T. Bower, Director of Human Resources, Wolfe, Inc.
From: Keith Madsen, M.D.
Date: April 1, 1998
Re: Patient 484-82-5553, Jack Adams

As per your request, this statement should serve as a summary of the condition of Jack Adams for the purposes of your employee review process. Our files are confidential, but the patient has agreed to release this information to Wolfe, Inc. as the enclosed release form indicates.

On October 18 of last year, the patient, Jack Adams, was seen at our clinic for treatment of a dislocated shoulder. The attached summary lists the procedures and costs associated with diagnosis and treatment.

On December 3, the patient returned complaining of limited neck pain and mobility problems. He also noted periodic numbness in the right forearm and hand. Initial findings suggested inflammation and overuse following the trauma of the October injury. The patient was prescribed a daily treatment of 2400 milligrams of ibuprophen to reduce inflammation and was encouraged to minimize use of the arm for a period of one week.

On December 12, the patient returned to the clinic noting no improvement in mobility or sensation. After consultation with orthopedist, B. G. Treu, and neuromuscular specialist, Juval Siraj, the team determined the need for several diagnostic procedures including X rays, EMG, and nerve conduction. Upon the completion of these tests, the team determined it was likely that the patient was suffering from radial neuropathy caused, at least theoretically, by a contusion or multiple contusions to the radial nerve. Without surgery, we could not determine the cause or extent of the damage or offer a prognosis for recovery. Although the team agreed that drug therapy might sustain the patient and delay surgery, it was unlikely to negate the need for surgery altogether. The patient determined to delay surgery and receive drug treatment and physical therapy for the period immediately following the December 12–15 meetings with the team.

On January 24 of this year, the patient was seen again and complained of even more loss of sensation in the forearm area. The team recommended exploratory surgery to assess extent of damage and repair initial findings. The surgery was scheduled for February 14. The first surgery indicated some blood infiltration into tissues caused by a radial nerve contusion. Such a contusion is usually caused by significant trauma. Scarring was indicated in two sections of the radial nerve. It was not possible to determine whether the accident in October of last year was the sole cause of the trauma and resulting scarring. Xrays also indicated some trauma to the brachial plexus in the axilla which may or may not be related to the nerve damage.

After the February 14 surgery, the team scheduled a second laser surgery to repair the scarring damage and transfer the nerve into muscle tissue for the purpose of stabilization. This surgery was scheduled for March 27. The second surgery was successful and the patient is currently recovering well.

Because neurosurgery is difficult and each patient responds differently, there are no guarantees that Mr. Adams will not suffer some further limitations. However, Mr. Adams is otherwise in fine condition and indicates a strong desire to recover and return to his professional responsibilities at Wolfe, Inc. Both of those conditions suggest favorable odds for his recovery.

I am happy to speak with you further if you require more insight into this case. Please feel free to contact me at the office if I may be of further service.

cc: B.G. Treu, J. Siraj, J. Adams

# COLORADO FAMILY CLINIC & HOSPITAL

CF✚CH

| DATE OF SERVICE | SERVICE CODE | CHARGE DESCRIPTION | QUANTITY | UNIT PRICE | AMOUNT | COMMENTS |
|---|---|---|---|---|---|---|
| 10/18 | 3270004 | DR FEE-LEVEL IV | 1 | 108.70 | 108.70 | 45 |
| 10/18 | 4070723 | STADOL INJ | 3 | 20.23 | 60.69 | 816 |
| 10/18 | 4070695 | PHENERGAN INJ | 1 | 12.50 | 12.50 | 816 |
| 10/18 | 3159076 | C-F PRIMARY Y SITES | 1 | 24.00 | 24.00 | 59 |
| 10/18 | 3159016 | LR 1000ML | 1 | 32.00 | 32.00 | 816 |
| 10/18 | 4012101 | CBC | 1 | 32.50 | 32.50 | 730 |
| 10/18 | 4011030 | POTASSIUM | 1 | 15.00 | 15.00 | 730 |
| 10/18 | 4010011 | BLOOD ALCOHOL/MED | 1 | 35.00 | 35.00 | 730 |
| 10/18 | 4010022 | BLOOD COLLECTION | 1 | 8.50 | 8.50 | 735 |
| 10/18 | 4040168 | SHOULDER 2 V | 1 | 70.00 | 70.00 | 709 |
| 10/18 | 4040167 | SHOULDER 1 V | 1 | 58.00 | 58.00 | 704 |
| 10/18 | 4040168 | SHOULDER 2 V | 1 | | 70.00- | 704 |
| 10/18 | 3250303 | NEEDLE IV CATH | 1 | 2.50 | 2.50 | 59 |
| 10/18 | 3231007 | ER LEVEL III | 1 | 75.00 | 75.00 | 59 |
| 10/18 | 3231003 | ER EXT SERV/15MIN | 2 | 12.00 | 24.00 | 59 |
| 10/18 | 3232007 | ADVANCED AMB SUPP 2 | 1 | 285.00 | 285.00 | 47 |
| 10/18 | 3232006 | AMBULANCE MILEAGE | 41 | 5.60 | 229.60 | 47 |
| 10/18 | 4040058 | DUPLICATION FILM | 1 | 7.00 | 7.00 | 704 |
| 10/18 | 3230007 | MED THERAPY-EMERG. | 1 | 50.00 | 50.00 | 59 |
| | | -- TOTAL AMOUNT DUE -- | | | 1,059.99 | |
| | | -- SUMMARY OF CHARGES -- | | | | |
| | | EMERGENCY PHYS | 981 | | 108.70 | |
| | | MEDICATION THPY | 260 | | 50.00 | |
| | | CONTRAST/SUPPLY | 621 | | 7.00 | |
| | | STERILE SUPPLIES | 272 | | 2.50 | |
| | | PHARMACY | 250 | | 73.19 | |
| | | IV'S & SOLUTIONS | 258 | | 56.00 | |
| | | LABORATORY | 300 | | 91.00 | |
| | | X-RAY TECH FEE | 320 | | 58.00 | |
| | | ER/OUTPT SERV. | 450 | | 99.00 | |
| | | AMBULANCE | 540 | | 514.60 | DAYS: 42 |

**TO ASSIST IN PAYMENT WE ACCEPT:** MasterCard VISA

**Please fill in the information below**

Bankcard Account Number ☐ Mastercard (✓) ☐ Visa ____ Expiration Date ____

1,059.99
UNPAID BALANCE

Cardholder Signature ______________ Amount $________ Date________

FORM NO.

# Summary of Evaluative Criteria for Major Tasks in Case 20

| | ***1 Unacceptable***<br>Insufficient answer to assignment expectations | ***2 Below Average***<br>Inappropriate or ineffective verbal/ visual choices limit document success | ***3 Meets Task Expectations***<br>Has answered objectives of assignment, but individual components could be strengthened | ***4 Above Average***<br>Few flaws, document meets expectations, but could benefit from more attention to detail | ***5 Excellent/ Professional***<br>Few or no flaws, demonstrates keen insight into case subtleties and details |
|---|---|---|---|---|---|
| **Purpose/Key Points**<br>▪ Identifies and meets purpose<br>▪ Articulates key points clearly and with attention to office dynamics | | | | | |
| **Context**<br>▪ Identifies/defines context and situational constraints<br>▪ Demonstrates awareness of the Adams situation<br>▪ Demonstrates understanding of technical concepts and terms | | | | | |
| **Audience**<br>▪ Identifies/defines audience and meets identifiable needs<br>▪ Establishes appropriate tone<br>▪ Understands technical details enough to communicate effectively | | | | | |
| **Synthesis**<br>▪ Demonstrates an ability to synthesize and appropriately employ information from a variety of sources | | | | | |
| **Design**<br>▪ Demonstrates awareness of visual design elements of task<br>▪ Demonstrates an awareness of design options & technological aids | | | | | |